AF589455

Textbook of
Environmental Sciences

The Authors

Dr. Divya Agarwal is **Ph.D.** in Environmental Sciences, BBA (Central) University, **gold-medalist** in the university merit in M.Sc. Environmental Sciences, qualified UGC-NET (2002, 2003, 2004), ICAR-NET (2001), various specialized trainings, course modules in Disaster Management, have publications in varied aspects of Environmental Sciences. She has been teaching as Associate Professor in Accurate Institute of Management and Technology (AIMT), Greater Noida since Aug, 2009. Earlier she was in Krishna Institute of Engineering and Technology (KIET). She carries teaching experience of graduate and post-graduate classes of more than eleven years. She has been a reviewer in Pearson Education Pvt. Ltd. and Journals of Elsevier Publications.

Dr. Manoj K. Agarwal, is **Professor (Chemistry)** and carries rich academic, research and administrative experience. Presently he is Director (Planning) in Accurate Institute of Management and Technology, Greater Noida. He is a co-editor in the Journal of Pure and Applied Sciences and Technology. Earlier, he was Director in DJ College of Engineering and Technology. He keeps on adding to his qualifications, showing the interest in academics and dynamic personality.

Textbook of *Environmental Sciences*

— Authors —
Divya Agarwal
Manoj K. Agarwal

2025
Daya Publishing House®
A Division of
Astral International Pvt. Ltd.
New Delhi – 110 002

First Publishied, 2015
Reprinted, 2025
ISBN 978-93-59194-88-2

Published by : **Daya Publishing House®**
A Division of
Astral International Pvt. Ltd.
– ISO 9001:2015 Certified Company –
4736/23, Ansari Road, Darya Ganj
New Delhi-110 002
Ph. 011-43549197, 23278134
E-mail: info@astralint.com
Website: www.astralint.com

Acknowledgements

Authors are grateful to the respected Chairman Sir, Mr. C.L. Sharma and the Jubilant Group Director Ms. Poonam Sharma for promoting an academic environment in the institute.

D. Agarwal is grateful to the parents to up-bring her with a positive attitude, inspite of all the adversities. Especially to the loving father Late Dr. Anand Prakash Agrawal, who encouraged to start up this project. It was his passion to develop her in academics. He was a Chemistry teacher in Kendriya Vidayalaya. Mother also provided constant encouragement. Mother-in-law Mrs. Mithlesh Jindal also provided great appreciation. Her husband, Mr. Neeraj Jindal, bhiaya and Bhabhi, Brother-in-law, sister-in-law and kids feel proud when I used to show them the manuscript. The small loving daughter, Samishta Jindal also keeps patience and provides great cooperation. Our landlords, Mrs. and Mr. Surekha Sharma are just like one more parents God has gifted.

M.K. Agarwal is grateful to his mother Smt. Kamal Agarwal; father Mr. G.C. Agarwal for the constant source of blessings. He is always thankful to his wife Ms. Shilpa Agarwal for inspiration. He is grateful to the children Ms. Vageesha Agarwal and Pranav Agarwal for love and affection.

Authors are highly thankful to Prof. Mohammad Yunus Ex-Dean, Department of Environmental Sciences, Babasaheb Bhimrao Ambedkar (A Central) University, Lucknow; Prof. D.P. Singh Department of Environmental Sciences, BBAU, Lucknow, Dr. Ram Boojh Environment Coordinator (UNESCO) and Mr. Paritosh Chandra Tyagi, Ex-chairperson, CPCB for their with their thoughts. Authors are thankful to Dr. S.K. Dubey, Director, Dr. Sandeep Tiwari, Director Research. Dr. Atul Choudhary, Deputy Director and Dr. Pramod Sharma, Dean Student's Welfare for promoting academics.

Jitendra, the peon of the second floor is the key person who provides cool drinking water again and again. *Water is the vital natural resource, important for life*!

Special thanks are due to Manish Kaushik, Assistant Professor who provided great help while discussions and providing helps as and when required. Thanks is also due for HOD Computer Science Department, Mr. Sansar Singh and all the concerned members for providing technical support of computers in the laboratory. Mr. Sahni, HOD Mechanics also provided encouragement. Ms. Pooja Yadav, Anupam Ratn, Dr. Anoop Singh, Dr. S.L. Rajpoot, Dr. Diwakar Chauhan, Mr. Amit Singhal, Dr. Ananta Srivastava and Ms. Manisha Mishra also deserve great thanks by providing a constant encouragement, and understanding. Mr. Rohan, Rishi Kumar and other faculty members keep motivating the first author by appreciating the contents and presentation in the notes provided to the students. Aditi and the colleague receptionists give great pleasure while wishing good-morning every day. We deeply acknowledge to all the colleagues, who appreciate our hard-work to grow in academics. Members of Registrar office also provided technical support.

We would be highly thankful to the readers who could provide their feedback (email: divya.sustainable@gmail.com). We are thankful to our dear students who initiated us to write the book and provided important feedback about the key features, which they need in a book. They always appreciate the notes provided to them and request to write a book. The authors are thankful to the entire global network of environmental sciences, which played useful role in systematic collection of thoughts in this book. Authors feel highly thankful to the entire team of the esteemed publishing house for their hard-work.

Dr. Divya Agarwal

Dr. Manoj K. Agarwal

Foreword

I like my struggles beginning last decade while implementing compulsory UGC module on environmental studies for all UG courses as per Hon'ble Supreme Court's directive in more than 70 Academic departments and affiliated colleges of Bundelkhand University as Director of Institute of Environment and Development Studies. Since then I have seen increasing recognition of environmental studies within engineering, business and management education as well. This 15 years period also saw the diversification of environmental studies and introduction specializations within, for example, on natural resources, EIA, disaster management, urban environment, climate change, etc. However, areas like environmental-health, system ecology, water sanitation – housing habitat and urban ecology are yet to be adequately addressed.

Environmental science, unlike many basic sciences, is the "science and management of change and relations" keeping context of 'eco' as forming both terms – "ecology" and "economics", is evolving day by day, is dynamic, and is thus, as explained by Eugene P. Odum (1980), a blend of hard and soft science - natural social and engineering disciplines. But, its constitution is "interdisciplinary" rather than "multidisciplinary" unlike medical or engineering sciences.

This book "*Textbook of Environmental Science*" authored by Dr. Divya Agarwal and Dr. M. K. Agarwal, who are known to me for long time and to the academic world for their contribution to environmental and technical education, has benefits over the existing titles due to its coverage of latest issues and update on developments. Authors long research and teaching experience is visible through the quality of the content which is with simpler language and understandability, is my conviction.

Writing this foreword is a pleasure, as I just was in a long telecon by United Nations Environment Programme (Geneva) regarding how environmental studies in Indian Universities grows and diversifies with latest issues emerging, in particular of focus on EIA/SEA, environmental audit, and environmental planning to include focus on climate change adaptation, environmental education and sustainability issues. I also foresee the MHRD/UGC intervention and establishment of dedicated Universities on environment and disaster studies, and revision of the UGC curriculum of JRF-NET, post-graduation and UG modules.

This book shall enable great help to the teachers and students of environmental science in Universities and engineering and management institutions across the country, I anticipate and am sure of. I congratulate the authors for presenting this value added textbook on the subject.

Dr. Anil Kumar Gupta

Sr. Associate Professor of Policy Planning and Environment, and
Director, Indo-German Cooperation Programme ekDRM,
National Institute of Disaster Management
(Ministry of Home Affairs, Govt. of India), New Delhi

Contents

Acknowledgements **v**

Foreword ***vii***

1. Environment, Environmental Sciences and Ecology **1**

1.1 Environment *1*

1.2 Components of Environment *1*

1.2.1 Atmosphere (Air) *1*

1.2.2 Hydrosphere *4*

1.2.3 Lithosphere *4*

1.2.4 Biosphere *4*

1.3 Environmental Sciences *4*

1.3.1 Multidisciplinary Nature of Environmental Sciences *5*

1.3.2 Scope of Environmental Sciences *5*

1.3.3 Importance of Environmental Sciences *7*

1.4 Ecosystem *8*

1.4.1 Definition of Ecosystem *9*

1.4.2 Concept of Balanced Ecosystem *9*

1.4.3 Biosphere *9*

1.4.4 Ecology *12*

1.4.4.1 Understanding Ecology *12*

1.4.5 Characterization of Ecosystems 12

1.4.5.1 Structure of Ecosystem 12

1.4.5.2 Food-Chains and Food-Web 17

1.4.5.3 Types of Ecosystem 17

1.4.6 Functions of an Ecosystem 18

1.4.7. Biogeochemical Cycling/Nutrient Cycling 31

1.4.7.1 Sulphur Cycle 31

1.4.7.2 Carbon Cycle 33

1.4.7.3 Nitrogen Cycle 33

1.4.8 The Y-Shaped Energy Flow Model 36

1.4.9 Ecological Pyramids 36

1.4.10 Ecorestoration of Damaged Ecosystems 39

1.4.10.1 Concept of Ecorestoration 40

1.4.10.2 Secondary Succession Forms the Basis of Natural Ecosystem Restoration 40

1.4.10.3 Ecorestoration Technology 40

1.4.10.4 Bioremediation as a Viable Tool for Ecosystem Restoration 41

1.4.10.4.1 Phytoremediation 41

1.4.10.4.2 Restoration of Wasteland by Phytoremediation 42

1.4.10.4.3 Restoration of Aquatic Ecosystems by Phytoremediation 43

1.4.10.6 Restoration of Marine Ecosystems: Microbial Remediation 44

1.4.10.4.7 Vermiculture and Vermicomposting 44

2. Effect of Human Activities on Environment **49**

2.1 Food, Shelter, Economic and Social Security 49

2.1.1 Food Security or Sustainable Food Supply System 49

2.1.2 Challenges to Food Security 50

2.1.3 Shelter Security 51

2.1.4 Economic Security 51

2.1.5 Social Security 51

2.2 Effect of Human Activities on Environment: Agriculture, Housing, Industry, Mining and Transportation Activities 51

2.2.1 Effect of Modern Agriculture on Environment 52

2.2.2 Effect of Housing on Environment 55

2.2.3 *Effect of Power Generation on Environment* 57

2.2.3.1 *Effect of River Valley Projects (Water Resource Projects) on Environment* 57

2.2.4 *Effect of Mining on Environment* 57

2.2.5 *Effect of Transportation Activities on Environment* 58

2.2.6 *Effect of Tourism on Environment* 58

3. Natural Resources 61

3.1 *Types of Natural Resources* 61

3.2 *Natural Resources Availability and Problems or Natural Resource Crisis and Population Explosion* 61

3.3 *Water as a Vital Natural Resource* 62

3.3.1 *Water Cycle (Hydrological Cycle)* 62

3.3.2 *Availability and Quality Aspects (groundwater depletion)* 63

3.3.3 *Water-borne and Water-induced Diseases* 63

3.3.4 *Fluoride Problem in Drinking Water* 64

3.3.5 *Arsenic Problem in Drinking Water* 65

3.4 *Minerals as a Natural Resource* 66

3.5 *Forest as a Natural Resource* 66

3.5.1 *Over-Exploitation of Forest Resources* 66

3.5.2 *Conservation of Forest Resources* 67

3.6 *Food Resources* 67

3.6.1 *Conservation of Food Resources: Sustainable Agriculture (Organic-Farming) or Solution to Modern Agriculture* 67

3.7 *Land as an Important Natural Resource* 69

3.7.1 *Land Degradation* 69

3.7.2 *Wasteland Reclamation (Restoration)* 70

4. Biodiversity 74

4.1 *Definition of Biodiversity* 74

4.2 *Hierarchical Classification/Origin of Biodiversity* 76

4.3 *Significance of Genetic Biodiversity* 79

4.4 *Measurements or Measures of Biodiversity* 80

4.5 *Taxonomic Classification of Biodiversity* 80

4.6 *India as a Mega-diversity Nation or Biogeographic Zones of India* 81

4.7 *Hotspots of Biodiversity* 84

4.8 *Threats to Biodiversity/How Man-made Extinction of Biodiversity is Caused?* 87

4.9 *Endemic Species of India* 89

4.10 *Endangered Species of India* 89

4.11 *Threatened Species of India (Eastern Himalayas)* 90

4.12 *Conservation of Biodiversity or Biodiversity Management* 90

4.12.1 *Types of Conservation of Biodiversity* 91

4.12.2 *Conservation Tools (Techniques) of Biodiversity* 93

4.13 *International Efforts to Save Biodiversity* 93

4.14 *Values (or uses or importance or benefits) of Biodiversity* 95

5. Energy Resources **100**

5.0 *Daily Life Examples: Save Energy Cartoon Series* 100

5.1 *Population Explosion* 100

5.2 *Classification of Energy Resources* 103

5.3 *Major Differences between Conventional and Non-conventional Energy Resources* 103

5.3.1 *Renewable (Non-conventional) Energy Resources* 104

5.3.1.1 *Solar Energy Applications* 105

5.3.1.2 *Hydro Energy* 108

5.3.1.3 *Wind Energy* 109

5.3.1.4 *Marine Energy or Ocean Energy* 109

5.3.1.5 *Tidal Energy* 110

5.3.1.6 *Geothermal Energy* 110

5.3.2 *Alternate Energy Sources* 110

5.3.2.1 *Hydrogen as a Fuel for Future or Hydrogen as an Alternative Energy Source* 110

5.3.2.2 *Microbial Fuel Cell* 112

5.3.2.3 *Energy Context with Respect to Indian Scenario* 112

5.3.2.4 *Energy Plantation* 113

5.3.3 *Conventional (Non-Renewable) Energy Resources* 113

5.3.3.1 *Fossil Fuels or Fossil Fuel Based Energy* 113

5.3.3.2 *Nuclear Energy* 117

5.4 *Steps of Government to Conserve Energy* 118

6. Sustainable Development **121**

6.1 Sustainable Development 121

6.2 Principles of sustainability 122

6.3 Measurement of Sustainability or Sustainable Ethics or Equitable Utilisation of Natural Resources 123

6.4 Sustainable Lifestyle (Role of an individual in sustainable development) 124

6.5 Challenges to Sustainable Development 124

6.6 International Efforts to Achieve Sustainability 127

7. Environmental Pollution **131**

7.1 Environmental Pollution 131

7.1.1 Air Pollution 131

7.1.1.1 Sources of Air Pollution 131

7.1.1.2 Effects of Air Pollution 132

7.1.1.3 Classification of Air Pollutant 132

7.1.1.4 Control Measures of Air Pollution 133

7.1.1.5 Air Pollution Disasters 134

7.1.1.6 Long Range Transport of Gaseous Air Pollutants 136

7.1.1.7 National Ambient Air Quality Standards 136

7.1.2 Water Pollution 139

7.1.2.1 Sources of Water Pollution 139

7.1.2.2 Types of Water Pollutants 139

7.1.2.3 Effects of Water Pollution 140

7.1.2.4 Water Quality Standards 140

7.1.2.5 Control of Water Pollution 140

7.1.3 Thermal Pollution 142

7.1.3.1 Sources of Thermal Pollution 142

7.1.3.2 Effects of Thermal Pollution 142

7.1.3.3 Control of Thermal Pollution 143

7.1.4 Waste Water Treatment Plant 143

7.1.5 Soil Pollution or Land Degradation 145

7.1.5.1 Sources of Soil Pollution 146

7.1.5.2 Effects of Soil Pollution 146

7.1.5.3 Control Measures 146

7.1.6 Solid Waste 146

7.1.6.1 Sources of Solid Waste 146

7.1.6.3 Effect of Solid Waste 147

7.1.6.3 Control Measures of Solid Waste 148

7.1.7 Status of Solid Waste Management in India 150

7.1.8 Hazardous Waste Management (HWM) 150

7.1.9 Noise Pollution 153

7.2 Air-borne Diseases 155

7.3 Toxic Substances: Toxicant, Toxicity and Toxicology 156

7.3.1 Factors affecting toxicity 156

7.4 Carcinogens 158

8. Environmental Impact Assessment **162**

8.1 Environmental Impact Assessment (EIA) 162

8.2 Government Body which Executes EIA in India 162

8.3 Environmental Effects Analysed under EIA 162

8.4 Process of EIA 163

8.5 EIA Ruling 1984 163

9. Environmental Laws **167**

9.1 Environmental Laws: Provisions in the Indian Constitution towards Environmental Protection 167

*9.1.1 Salient Features of Air (Prevention and Control of Pollution) Act, 1981** 168

9.1.1.1 Amendments 169

9.1.1.2 Concluding Remarks of Air (Prevention and Control of Pollution) Act, 1981 169

*9.1.2 Salient Features of Water (Prevention and Control of Pollution) Act, 1974** 169

9.1.2.1 Concluding Remarks of Water (Prevention and Control of Pollution) Act, 1974 171

*9.1.3 Salient features of Forest Conservation Act, 1980** 171

*9.1.4 Salient Features of Wildlife Protection Act, 1972** 172

9.1.5 Salient Features of Environment (Protection) Act, 1986 172

9.1.5.1 Amendments in EPA 175

9.2 Role of Government in Environmental Protection 175

10. Global Environmental Issues **178**

10.1 *Green House Effect and Global Warming* 178

10.2 *Global Climate Change* 180

10.2.1 *International Efforts to Control Global Warming or Global Climate Change* 181

10.3 *Ozone Layer Depletion or Ozone Hole* 181

10.4 *Acid Rain* 182

10.5 *El Nino* 182

10.6 *La Nina* 182

10.7 *Automobile Pollution* 182

10.8 *Urbanization and Environment* 183

10.9 *Population Growth* 184

11. Environmental Education **187**

11.1 *Environmental Education* 187

11.2 *Principles of Environmental Education* 187

11.3 *Need for Public Awareness or Importance of Environmental Education: Active public participation* 188

11.3.1 *Details of Environmental Movements Indicating the Importance of Environmental Awareness* 188

11.3.1.1 *Chipko Movement in Tehri Garhwal Gained International Importance* 188

11.4 *The 'Guidelines for Excellence' for Environmental Education* 189

11.5 *Process of Public Awareness* 190

11.6 *Agencies Active in Creating Public Awareness (Environmental Education): Non-Governmental Organizations* 190

11.7 *Conferences, Workshops, Seminars* 193

12. Social Environmental Issues **197**

12.1 *Social Environmental Issues are Important for Sustainable Developmental Strategies* 197

12.1.1 *Watershed Management* 197

12.1.1.1 *Rainwater Harvesting* 198

12.2 *Flyash Landfill Reclamation* 198

12.3 *Disaster Management* 198

12.3.1 *Urban Disasters* 198

12.3.1.1 *Flood* 199

12.3.1.2 *Cyclone* 200
12.3.1.3 *Landslides* 200
12.3.1.4 *Earthquakes* 200
12.3.1.5 *Drought* 201
12.3.1.6 *Hazardous Chemicals* 201
12.4 *Animal Husbandry* 202
12.5 *Women Education* 202
12.6 *Environmental Problems in India (Developing Countries)* 203

13. Human Population **207**
13.1 *Human Population* 207
13.2 *Population Growth Rate* 207
13.2.1 *Factors Affecting the Growth Rate* 207
13.2.2 *Population Explosion* 208
13.2.3 *Causes of Population Explosion* 208
13.2.4 *Effects of Population Explosion* 208
13.2.5 *Control Measures of Population Explosion* 208
13.3 *National Population Policy, 2000 (NPP, 2000)* 209
13.4 *Environment and Human Health (UPTU, 2007, 2009)* 209
13.4.1 *Sexually Transmitted Diseases (STD)* 209
13.4.1.1 *HIV/AIDS* 209
13.5 *Role of IT in Environment and Human Health* 209
13.6 *Human Rights* 209
13.7 *Value Education* 209

Previous Papers **211**
Glossary **231**
Index **243**

Chapter 1

Environment, Environmental Sciences and Ecology

1.1 Environment

Environment is our surroundings which includes living and non-living factors and their interactions with each other. It is also defined as the sum total of all social, economical, biological, physical or chemical factors. It is the surroundings of man.

Word *environment* is derived from a *french* word *environ* or *enviroonner* meaning *"to encircle". Environment includes all the physical and biological surroundings of organisms along with their interactions"* **(Environmental Protection Act, 1986).**

1.2 Components of Environment (*UPTU, 2012-13*)

The biosphere constitutes three components including atmosphere, lithosphere and hydrosphere. Air envelop around us *i.e.* our atmosphere is responsible for maintaining temperature by protecting earth from excessive heat during day time and excessive cold during night time, and regulates the water cycle.

1.2.1 Atmosphere (Air)

Atmosphere is the gaseous cover, suspended solids and liquids around the earth surface. Atmosphere comprises of air. Air is a mixture of gases. Atmosphere extends up to a height of 1600 kilometres but living organisms are present only up to 6 kilometres.

Composition of Atmosphere

Many gases are present in atmosphere (**Table 1.1**). Out of these, O_2 and CO_2 are essential for existence of life and make the earth as an only planet in the solar system, where life is possible.

Atmospheric Profile

Atmospheric profile is the arrangement of layers of atmosphere with respect to temperature and height. (*UPTU*, **2011-12, 2010**)

Atmosphere comprises of air or a thick blanket (=envelope) of air which surrounds our earth. Air is a mixture of gases. For example: nitrogen (78.03 per cent), oxygen (20.99 per cent), carbon-di-oxide (0.034 per cent), inert gases (0.95 per cent), water vapours and dust particles (variable). The earth's atmosphere is divided in to 4 different layers: *troposphere, stratosphere, mesosphere and thermosphere* (Table 1.2).

Table 1.1: Chemical Composition of Air.

Gas	*Per cent by Volume*
Nitrogen	78.084
Oxygen	20.9476
Argon	0.934
Carbon-di-oxide	0.0314
Neon	0.001818
Methane	0.0002

Source: A.K. De, "*Environmental Chemistry*".

(a) **Troposphere:** The closest layer of atmosphere to the earth surface is troposphere, where cloud formation occurs and air pollution spreads.

(b) **Stratosphere:** The ozone layer is present in stratosphere and protects earth's atmosphere from the sun's radiation of UV rays.

(c) **Mesosphere**: Meteorite burning takes place in the atmosphere.

(d) **Thermosphere**: All the gases move away. No activity can be done, due to very high temperature. It is divided into two layers:

  (i) Ionosphere

  (ii) Exosphere

Environmental Lapse Rate (ELR) or Lapse Rate

The decrease in temperature is observed with increasing height, known as Environmental Lapse Rate *i.e. Temperature is inversely proportional to height.* **Temperature decreases with height *i.e.***

$$\textbf{Height} \propto \frac{1}{\text{Temperature}} = \text{Positive environmental lapse rate.}$$

The negative lapse rate is called as temperature inversion.

Table 1.2: Atmospheric Profile.

Layers of Atmosphere	*Troposphere*	*Stratosphere*	*Mesosphere*	*Thermosphere*
Height (km)	(0) - (12)	(12)-(55)	(55)-(80)	(80) and above
Temperature (°C)	(17)-(-55)	(-55)-(0)	(0)-(-75)	(-75) and above
Environmental Lapse Rate (ELR)	+ ive	– ive	+ ive	– ive
Limiting Layer	Tropopause	Stratopause	Mesopause	Ionosphere extends upto 400 kms
Activities	Cloud formation, air pollution	Ozone layer formation	Meteorites-burning	Very high temperature

Source: Millers, "*Environmental Science*".

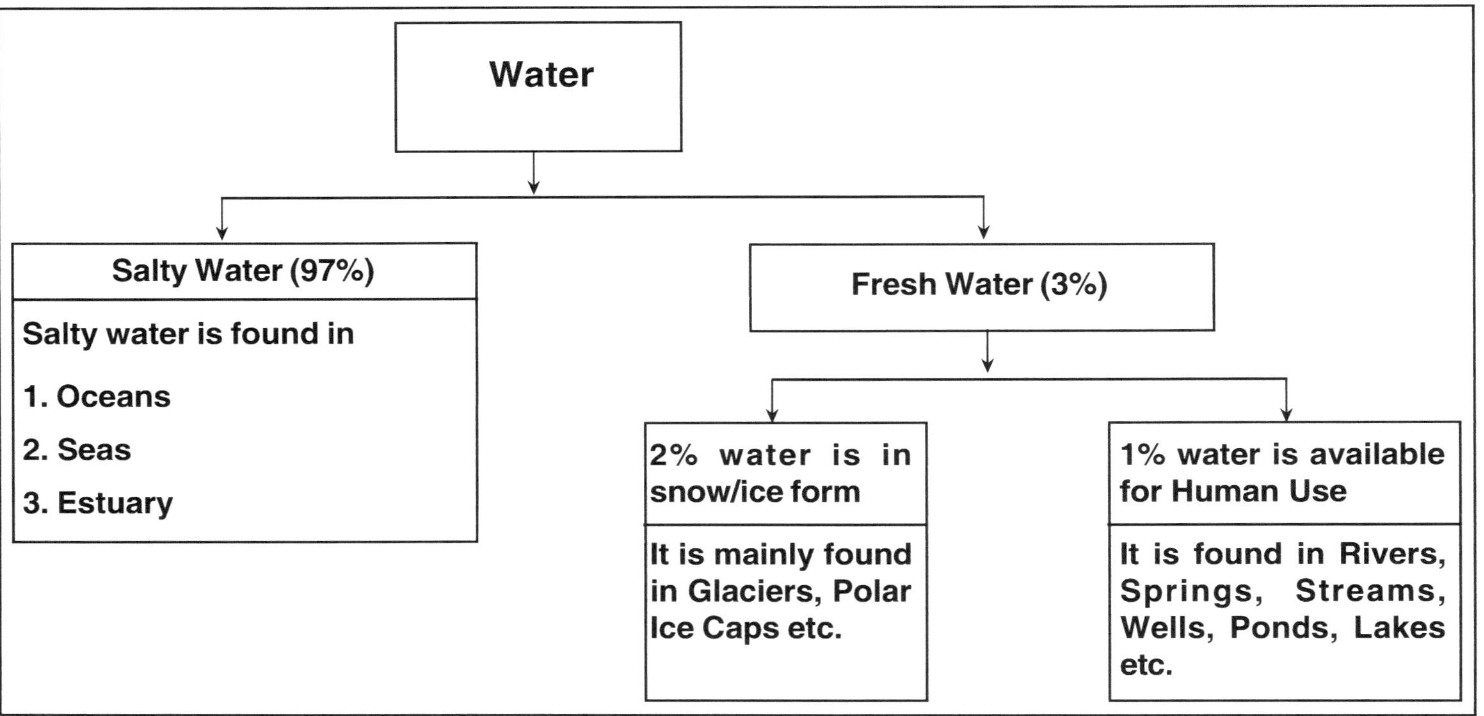

Figure 1.1: Distribution of Water in Hydrosphere.
(*Source*: Miller's Environmental Science)

1.2.2 Hydrosphere

Water is very important natural resource *i.e.* present in large amount in nature's reservoir and useful to man. *More than 70 per cent of the earth's surface is covered with water; therefore earth is called blue planet.* Of the total water available on the earth, 97 per cent water is salty and only 3 per cent water is fresh water. It is a universal solvent (Figure 1.1).

1.2.3 Lithosphere

Lithosphere comprises of soil and rocks component of the earth. Soil is the upper fertile layer of earth in which plants grow. Rocks are the solid part (=stones) of the earth's crust. Rocks are of three types: igneous, sedimentary and metamorphic.

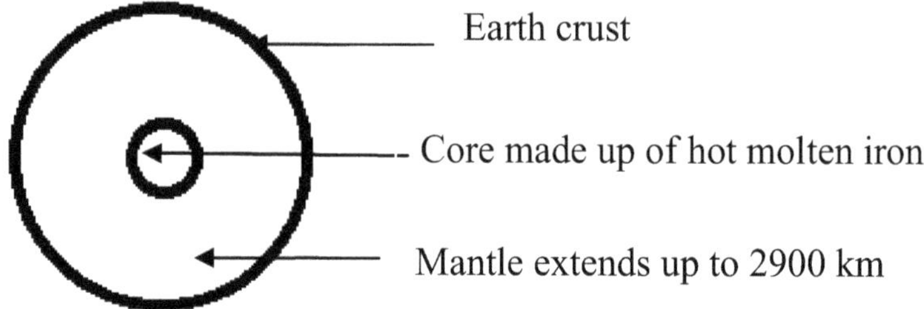

Figure 1.2: Internal Structure of Earth (Lithosphere).

1.2.4 Biosphere

Biosphere is the part of environment that supports life. It is the largest biological system on the earth and it consists of smaller functional units called ecosystems. Biosphere is also called ecosphere. Biosphere has two components: biotic components and abiotic components. Biotic components include plants, animals, human beings and micro-organisms. Abiotic components include air (=atmosphere), water (=hydrosphere) and soil (=lithosphere) (**Figure 1.3**).

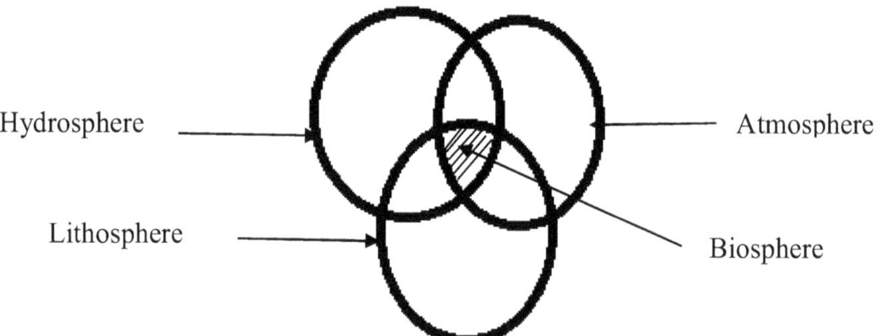

Figure 1.3: Formation of Biosphere.

1.3 Environmental Sciences

The systematic study of our environment *i.e.* interrelationship among the surrounding factors is known as environmental sciences.

1.3.1 Multidisciplinary Nature of Environmental Sciences

It is an integrated or multidisciplinary approach, with basic sciences, social sciences, physical sciences, chemical sciences, ecology, management, engineering and other professional disciplines (**Figure 1.4, Table 1.3**). Varied issues of environmental conservation and ecological importance are important in today's scenario of natural resource crisis and helps ensure sustainability, which is the ultimate goal of engineering curriculum (**Figure 1.5**).

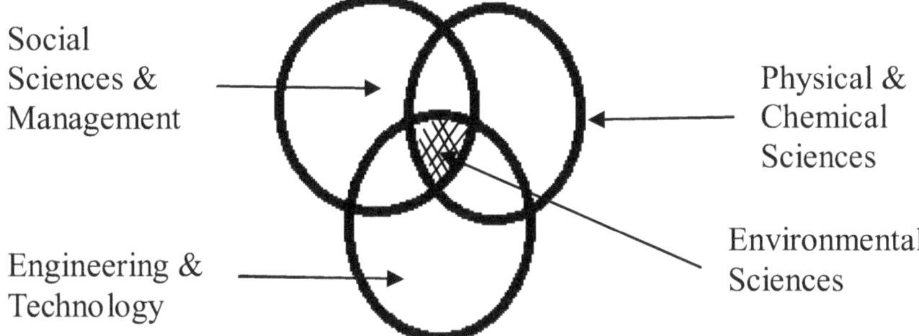

Figure 1.4: Environmental Sciences as a Multidisciplinary Approach.

Table 1.3: Finding Effects of Air Pollution will Involve Inputs.

Branch of Science	*Environmental Effects*
Botany	Effect of air pollution on plants.
Zoology	Effect of air pollution on animals.
Microbiology	Effect of air pollution on micro-organisms.
Ecology	Effect of air pollution on ecosystem.
Physics	Effect of air pollutants on atmosphere.

1.3.2 Scope of Environmental Sciences

Environmental Sciences is multi-disciplinary in nature. It gives knowledge of:

(i) Environmental Engineering

Branch of science involved in *control of environmental pollution* is known as environmental engineering *e.g.* **NEERI**: National Environmental Engineering Research Institute in Nagpur.

(ii) Environmental Management (Natural Resource Management)

Efficient use of natural resources (air, water, land, forest *etc.*) to ensure their sustainable utilization for present and future generation, is known as environmental management.

(iii) Biodiversity

Variety of living organisms and their habitats are known as biodiversity. Saving wild animals and plants from getting extinct, is an important part of environmental sciences.

(iv) Waste Management

Refuse, reduce, reuse, recycle, recovery, treatment and disposal are the important steps of waste management.

(v) Ecology

Study of ecosystem is known as ecology. The ecological interactions are governed by variety of phenomenon and studying them would involve various branches of science.

(vi) Global Environmental Problems

The long range transport of air pollutants is responsible for many environmental problems of global concern *e.g.* acid rain, ozone layer depletion, global warming and global climate change.

(vii) Environmental Laws

Laws are made to control environmental pollution *e.g.* Air (Prevention and Control of Pollution) Act 1981, Water (Prevention and Control of Pollution) Act 1974, Environmental Protection Act 1986 *etc.*

(viii) Social Environmental Issues

The issues of major concern are associated with our society *e.g.* women education, child welfare, poverty, population, pollution (3Ps), literacy, unemployment.

(ix) Environmental Sciences help achieve sustainability or Environmental Sciences is a tool to sustainable development

Applying the concepts of environmental sciences help students to grow future green engineers. *i.e.*

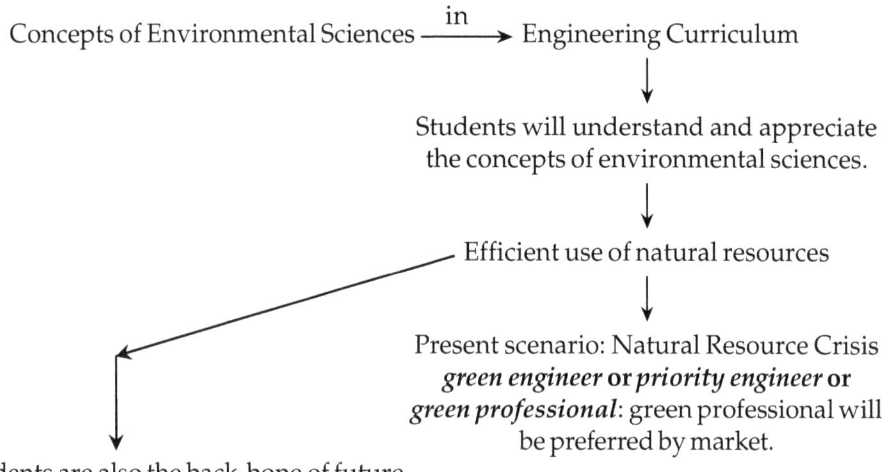

Figure 1.5: Environmental Sciences as a Tool to Sustainability.

Population increases exponentially, putting pressure on natural resources causing **natural resource crisis** known as Natural Resource Crunch (in the market). Learning environmental sciences will help students, understand the concepts and their applications. Thus, they will be the priority professionals, preferred by the market who could utilize natural resources efficiently known as **priority engineers** or **green engineers** or **priority professionals** (**Figure 1.5**).

The first and foremost thing is to understand the importance of environmental sciences in building the future economy for the coming generations. Economy of no nation can survive without being a developing ***'green-economy'***. It is not the chemistry which lets runs chemical reaction towards completion. It's the ***'green-chemistry'*** which facilitates complete combustion reaction, reduces waste generation, provides economic incentives to the production cell. Similarly it's not the professional, but the ***'green-professional'*** will be the priority of sustainable future who could utilize natural resources by being the most frugal engineer, a pharmacist, a microbiologist, doctor, real estate practitioner *etc.* Corporate are actively involved to formulate a ***'green-marketing'*** strategy because that helps attract consumer as well as government attention.

Environmental sciences help students develop an attitude towards appreciation of concepts of environmental services, which will automatically update into an active involvement as well providing students the basic updates of sciences and technology in environmental sciences *i.e.* the basic analytical tools.

The importance of environmental sciences should even be recognized even at higher degree programs. The important topics of utility of the subject are monitoring of air pollutants, control devices of air pollutants, pH measurements, BOD, COD analysis, and turbidity measurement. This is how environmental sciences can be taught in a systematic time-dependent frame of curriculum well designed separately for each of the selected group of graduates (science, commerce, arts, engineering, and pharmacists).

1.3.3 Importance of Environmental Sciences

The systematic study of our environment *i.e.* interrelationship among the surrounding factors is known as environmental sciences. It is an integrated or multi-disciplinary approach, with basic sciences, social sciences, physical sciences, chemical sciences, social sciences, ecology, management, engineering and other professional disciplines.

Students will understand and appreciate the concepts of environmental sciences.*viz.*

1. **Natural Resource Conservation**: Efficient use of natural resources
2. **Sustainable Development**: Integration of socio-economic and environmental benefits in developmental activities.
3. **Sustainable Agriculture**: Integration of traditional and modern agricultural techniques.
4. **Waste Management**: Efficient use of natural resources helps minimize waste generation.

5. **Ecology**: The study of ecosystem.
6. **Human interference in an Ecosystem**: Man-made activities mostly disturb the balance in an ecosystem.
7. **Green Chemistry**: Conducting chemical reactions towards nearly completion.
8. **Green Marketing**: Publicity of eco-friendly potential of a product.
9. **Biodiversity Conservation**: Efforts for protecting the species of plants and animals whose numbers are decreasing to an alarming extent.
10. **Environmental Law**: Legislation framed to control environmental pollution.
11. **Environmental Education**: Creating awareness among the local public about local environmental issues.
12. **Environmental Pollution**: Unwanted changes in the properties of air, water and soil, which adversely affects living systems.
13. **Environmental Engineering**: Branch of engineering to control environmental pollution.
14. **Environmental Management**: Utilization of natural resources, ensuring their conservation towards a sustainable future.
15. **Environmental Impact Assessment**: A continuous process in which environmental impacts of mega-developmental projects are minimized.
16. **Energy Conservation or Green Energy**: Efficient utilization of energy resources to conserve their potential for future.

"Environmental science is actually the continuous experiments of nature"

A deep insight to sources of environmental pollution would automatically indicate the courses or environmental pathways of pollutants *i.e.* effects of environmental pollution. Understanding the sources and effects of environmental pollution, would help minimize the adverse effects *i.e.* control measures. When students minimize the negative environmental impacts of developmental activity, we are approaching to **sustainable development**.

1.4 Ecosystem

Biotic (living) and abiotic (non-living) factors interact with each other to form a self-sufficient unit known as ecosystem. e.g. Pond, Desert, Forest, River ecosystem. The term ecosystem is coined by A.G. Tansley.

Concepts of Ecosystems

Ecosystems are interrelated natural systems constituting living organisms and their non-living environment. Organisms form the living or biotic component of an ecosystem. Variety of interactions exist among these various components. The study of interrelationship between the living and non-living components *i.e.* the web of mutual interactions is known as ecology.

1.4.1 Definition of Ecosystem

The term ecosystem was given by Sir Aurthur Tansley (1935). Living or biotic and non-living or abiotic factors interact with each other to form a self-sufficient unit known as an ecosystem. An ecosystem is also defined 'as a structural and functional unit of biosphere constituting living beings and the physical environment both interacting and exchanging materials between them.' (**Figure 1.6**).

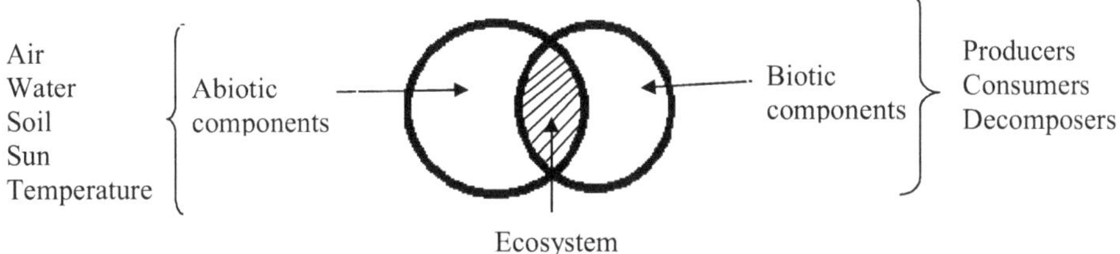

Figure 1.6: Components of an Ecosystem.

1.4.2 Concept of Balanced Ecosystem

Ecosystem is a unit of steady state equilibrium in the biosphere. The property of living organisms being in a steady state is known as **homeostasis**. In other words, homeostasis is the property of self-sufficiency of an ecosystem measured in terms of continuous energy flow and nutrient flow. Growth promoting factors *viz.* reproduction, adaptation, suitable climatic conditions and growth reduction factors *viz.* competition, disease, unsuitable climate *etc.* maintain a state of balance in an ecosystem known as homeostasis (**Figure 1.7**).

Living organisms exhibit a physiological range of stress within which they can tolerate its minimum and maximum values known as *range of tolerance* or *physiological range of tolerance*. Ecosystem has developed a system of sending its feedback to nature, regarding the range of tolerance known as feedback mechanisms. Beyond the upper and lower limit of stress, ecosystem sends these signals known as negative feedback mechanisms to bring the ecological system under ideal conditions *i.e.* homeostasis. If condition reverses and physiological stress further increases, then accelerating type of mechanisms operate known as *positive feedback mechanisms*, which may cause the system to collapse.

Thus human activities should help maintain homeostatic balance in an ecosystem and should not contribute to positive feedback mechanisms (**Figure 1.8**).

1.4.3 Biosphere

Biosphere is regarded as a global ecological system or global ecosystem. The ecosystem is a smallest unit with variety of interactions in living and non-living components taking place in such a way to maintain stability. Similarly complex series of interactions among the various components govern biogeochemical cycling and maintain the ecological balance in the biosphere.

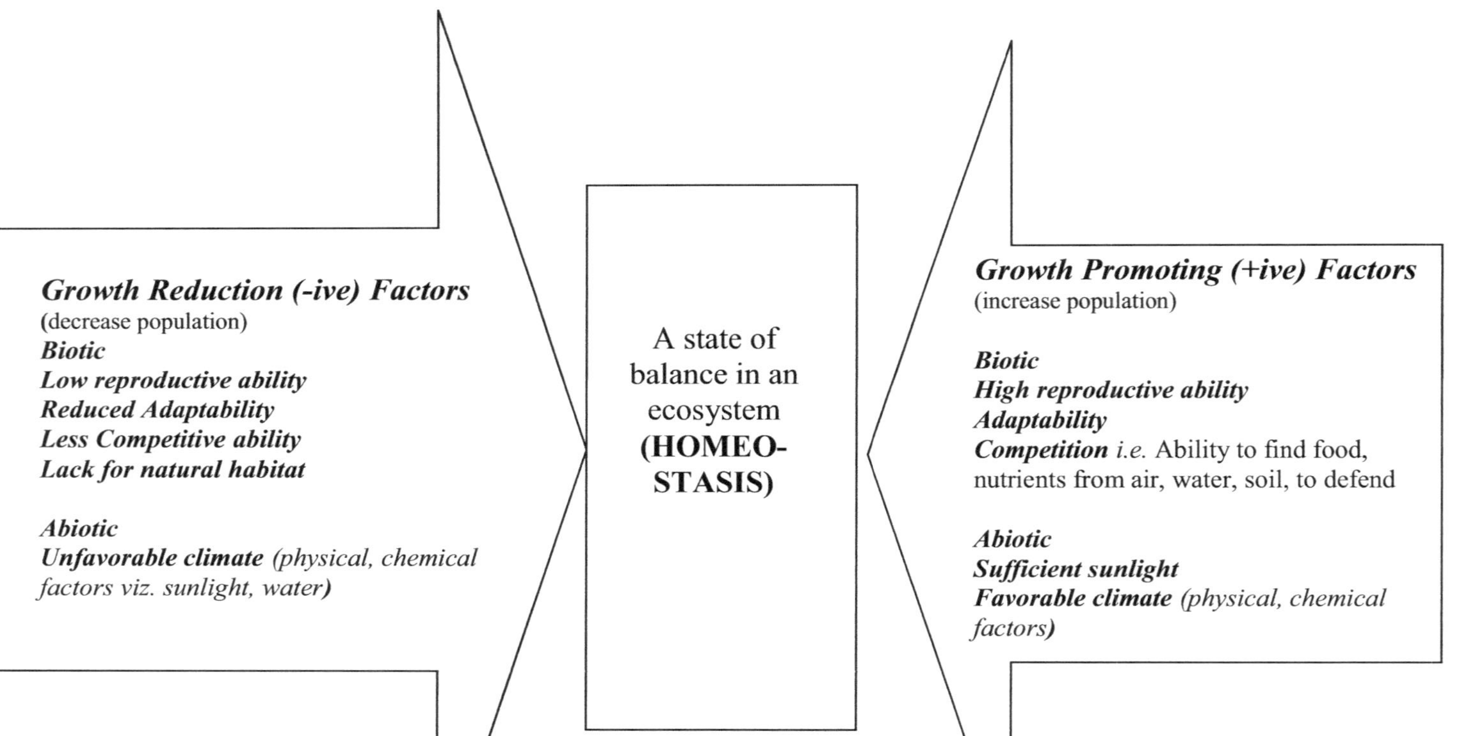

Figure 1.7: Maintaining Ecosystem Balance (Homeostasis) through Negative Feed-back mechanism.

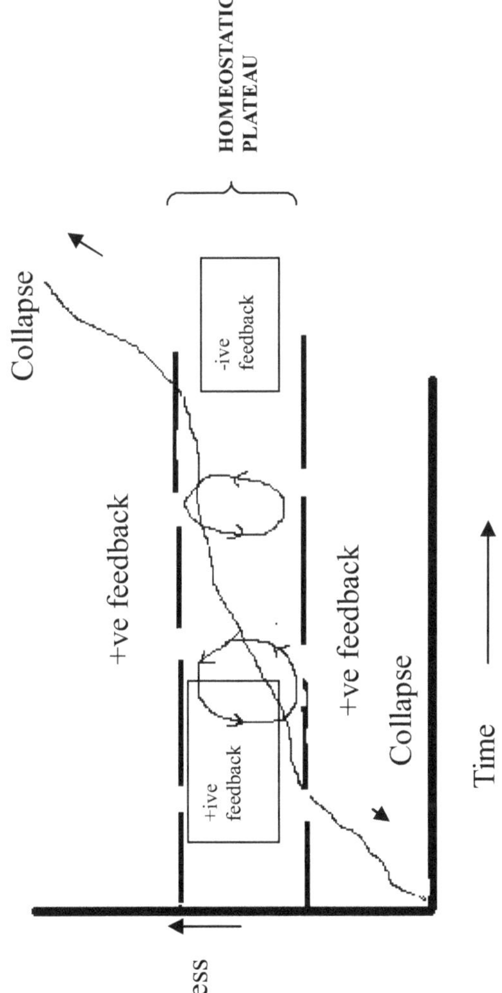

Figure 1.8: Ecosystem Balance by Positive and Negative Feedback Mechanisms.

1.4.4 Ecology

The study of interrelationship among the living and non-living factors is called as ecology. The term ecology (Gk. Oikos, house or place to live, logos-study or discourse) was given by Ernest Haeckal (1869) and defined it as "the science which deals with reciprocal relations of organisms and the external world" Elton Charles a british ecologist defined it as "a scientific natural history concerned with sociology and economics of animals." Warming (1895) stated ecology as the study of organisms in relation to their environment. Federick Clement said ecology as "the Science of community." E. Odum (1962) clearly stated as "study of structure and function of ecosystems." Macfadyen (1957) stated as "ecology concerns itself with the inter-relationships of living organisms and their environment." It can also be called as science of living environment.

The study of reciprocal relationships between organisms and their environment or the study of interrelationship between biotic and abiotic factors is known as ecology."

1.4.4.1 Understanding Ecology

Exploring our connections to the living world *i.e.* the ways in which human systems depend on natural systems and the ways in which humans affect them, is exploring ecology. Professor Garrett Hardin, a renowned ecologist wrote that ecology takes as its domain the entire living world. Ecological interactions are the most complex interrelated networking in nature. *e.g.* Lotka-volterra model is a famous ecosystem model which describes prey-predator relationship. So an insight towards functioning of nature would enable humans to develop harmony with nature and ecological phenomenon.

The term ecology is mis-used very often in literature. It been claimed many times "Save our ecology", ecology is in danger. Chipko movement is also incorrectly described as ecological movement. Ecology should not be taken synonymous with the word environment. Here ecology can be replaced by environment or more technically biosphere.

1.4.5 Characterization of Ecosystems

1.4.5.1 Structure of Ecosystem

An ecosystem, no matter how small, contains two components-the living or biotic components and non-living or abiotic components. Biotic components include autotrophs and hetrotrophs. Heterotrophs are of two types: consumers (herbivores and carnivores, omnivores) and decomposers (**Figure 1.9**).

Abiotic Components

The abiotic components of an ecosystem are the physical and chemical factors. Existing range of abiotic conditions decide survival of living organisms. Alterations in these conditions can have severe consequences. Single term for the long term combination of abiotic conditions is climatic factors: air, water, wind, temperature, humidity, rainfall, sunlight, mist, fog, water vapours and other physical factors. Following abiotic factors known as **standing stage** forms the **nutrient-pool** of nature in air, water and soil components.

(i) Inorganic Substances

Elements such as Hydrogen (H), Oxygen (O), Nitrogen (N), Sulphur (S), Phosphorous (P), Potassium (K), Calcium (Ca), Magnesium (Mg), Sodium (Na), Beryllium (Be), Boron (B), Helium (He) *etc.* are the important inorganic substances. Chlorophyll molecules help green plants carry out photosynthesis.

(ii) Organic Substances

Carbohydrates, proteins, lipids and humic acids form important components of organic substances. Factors related to soil are known as *edaphic factors.*

All the inorganic and organic substances keep circulating in the biogeochemical cycling or nutrient cycling from air, water and soil to the biotic components.

Importance of Abiotic Factors: Range of Tolerance

Organisms survive in a range of environmental conditions by undergoing series of adaptations known as a range of tolerance. Organisms survive to their best in the optimum range. At the outer margins of optimal range, zones of physiological stress exist where survival and reproduction are possible with difficulty *i.e.* hibernation, *etc.* Outside these zones, there are zones of intolerance, where the degree of hardness of abiotic factors leads to death of living organism. Variety of factors influence the degree of tolerance *i.e.* age, species, behaviour, natural events like floods, tropical storms and volcanic eruptions all change conditions for varying periods.

For example, incidence of deaths due to severe chilling increases in low temperature conditions indicating the zone of intolerance. Warmer temperatures represent zone of optimal growth. Little chilling is the zone of physiological stress.

Growth Limiting Factors

A combination of different abiotic factors is responsible for the growth of living organisms wherein one factor plays important and critical to the survival and growth of a population known as limiting factor. Changes in limiting factors can adversely effect in growth. For example In Tropical forest ecosystem, sunlight regulates the growth of tree species and the growth of small herbs within the canopy area of trees is adversely affected.

Biotic Factors

Living organisms are the biotic components of ecosystems. Bacteria, algae, plants, fungi and animals are the various organisms present in various ecosystems. A group of organisms of same species within an ecosystem is known as population and population of different organisms coexist in an interdependent biological community. Interactions among the living organisms are numerous studied in ecology.

Species composition, stratification and trophic structure or organizations are the structural features of biotic components of an ecosystem.

(i) Species Composition

A combination of species present in an ecosystem represents species composition of that ecosystem. For example: in a tropical rain forest, a large number of varieties of species is found in forest ecosystem where as a few species occur in a desert ecosystem.

Species is a term given to group of individuals who can freely inter-breed among themselves.

(ii) Stratification

Population of particular kind of species forms variety of layers or strata in each ecosystem. For example: In tropical rainforest, the crown of trees, bushes and ground vegetation form different strata and are occupied by different species. On the other hand, the desert ecosystem shows a discontinuous layer of bare patches of soil with scarce spine-like vegetation or thorny.

(iii) Trophic Levels

Group of organisms in an ecosystem with similar eating habits constitute the trophic levels in an ecosystem. The concept of trophic level was first given by Lindeman (1942) who states a food chain consists of various steps for example: producers, herbivores (=primary consumers), carnivores (=secondary and tertiary consumers). *Each step of a food chain is known as a trophic level. It is also known as food level or feeding level.* The basic/first trophic level (T1) of an ecosystem is producers or autotrophs. Herbivores form the second trophic level (T2) or they are the primary (1*) consumers. Carnivores are the secondary (T3, 2*) and tertiary (T4, 3*) consumers.

For example: Food-chain in grassland ecosystem represents following trophic levels:

Grass	→	**Rabbit**	→	**Fox**	→	**Lion**
(Producers)		(Herbivores)		(Small carnivores)		(Big carnivores)
(T1)		(T2)		(T3)		(T4)

(a) Producers (Autotrophs)

They are the chlorophyll (a green pigment) containing plants *i.e.* green plants, which carry out **photosynthesis** in the presence of CO_2, H_2O, sunlight and soil. Every food-chain starts with green plants. They are also known as autotrophs or **transducers**.

Producers of desert ecosystems are quite different than that of Tropical forest ecosystem. The abiotic conditions of desert are scarce rainfall, sandy soil with low water-holding capacity, high temperature, strong winds *etc.* Plants growing there need to adapt to these prevailing abiotic conditions *e.g.* reducing their leaf surface area (to reduce the rate of evaporation) to spines *i.e.* variety of interactions prevail between the living and non-living factors.

(b) Consumers

Herbivores (Primary Consumers)

Animals which eat green plants are known as herbivores. Grasshopper, cow, deer, rabbit, goat *etc.* are the examples of herbivores of terrestrial ecosystem. They are also called as **key industry animals** because they convert plant material into animal material.

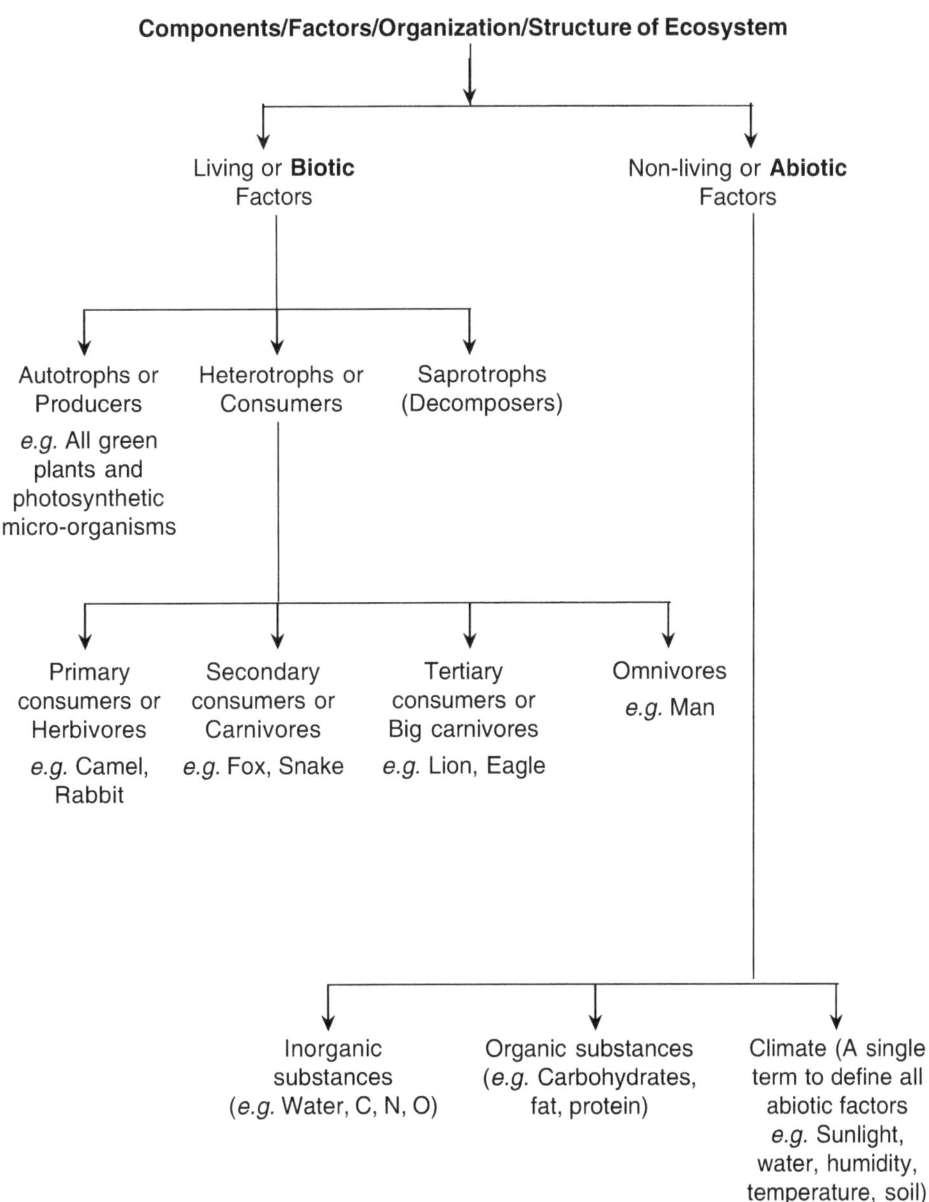

Figure 1.9: Structure (Components) of an Ecosystem.

Carnivores

Animals which feed upon herbivores are primary carnivores or second order consumers *e.g.* fox, cat, dog, birds. The animals which eat primary carnivores are known as secondary consumers or third order carnivores *e.g.* peacock, owl.

Some larger carnivores eat secondary consumers known as tertiary consumers. The top level carnivores which cannot be preyed upon, occupy top position in the food-chain are known as top-level carnivores.

(c) Decomposers

Organisms that breakdown complex organic matter to simple form, are known as decomposers *e.g.* insects, fungi, earthworms. They are also known as **micro-consumers** (very small sized consumers), **reducers** (decompose and remove dead matter) or **saprophytes or saprotrophs** (sapro = to decompose) or **detrivores** (which feed up dead and decaying matter) or **scavengers**.

The process of breaking down complex organic matter to simpler form, by secreting digestive enzymes in the surroundings (extracellular digestion), is known as mineralization. The released minerals are the raw material for green plants to grow and some of the decomposition products are absorbed by micro-organisms for their growth and metabolism.

Trophic Structure of Ecosystem

Living material present in different trophic levels of an ecosystem at a given time is known as standing crop. It is measured as number or biomass of organisms per unit area. The biomass is measured as per fresh weight or dry weight. More precise measurement of biomass is dry-weight because it reduces the variations in weight due to seasonal differences of water content (**Figure 1.10**).

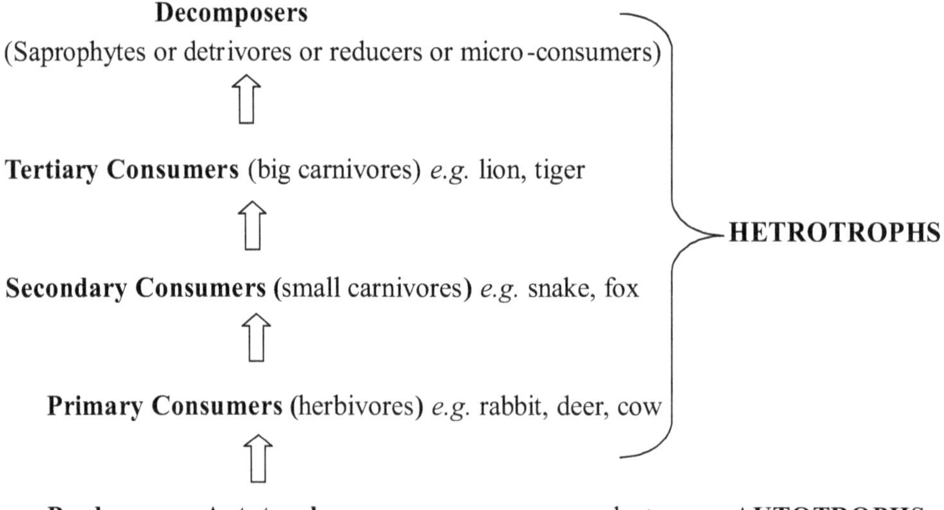

Figure 1.10: Trophic Structure of an Ecosystem.

1.4.5.2 Food-Chains and Food-Web

Food-chain

Transfer of food energy from one trophic level to another (*i.e.* from one group of organism to another) in an ecosystem. *e.g.* Grass → Rabbit → Lion. (Grassland ecosystem)

Types of Food-Chain

i) Grazing Food-Chain

It starts with green plants in a terrestrial ecosystem. *e.g.* Grass → Rabbit → Lion.

ii) Detritus Food-Chain

It starts with dead organic matter in an ecosystem *e.g.*

Mangrove Ecosystem

Fallen plant leaves → Insect larvae → Small fish → Big fish.

Food-Web

Inter-linkages (inter-networking) among the various food-chains in an ecosystem, is known as food-web. *e.g.* Food chains in grassland ecosystem are given as follows:

i) Grass → Grasshopper → Eagle
ii) Grass → Grasshopper → Lizard → Eagle.
iii) Grass → Rabbit → Eagle
iv) Grass → Mouse → Snake → Eagle

Food web of food-chains (i), (ii), (iii), (iv)

Thus, food-web describes complete feeding relationships in any described ecosystem. Nature provides variety of substitute foods via connecting the food chains.

Importance of Food-Web

Various organisms in the food web help maintain natures balance mechanism and provide various substitutes or alternatives in case of unavailability of one or the other organism due to human activities. For example application of rodenticide kills rodents *i.e.* rabbit from the grassland ecosystem, causing overgrowth of the grasses and extinction of the animal's dependent upon rabbits *i.e.* snakes, in case none of the other food-chains were present. This is how nature maintains a check-mechanism.

1.4.5.3 Types of Ecosystem

Showing in Figure 1.11.

1.4.6 Functions of an Ecosystem

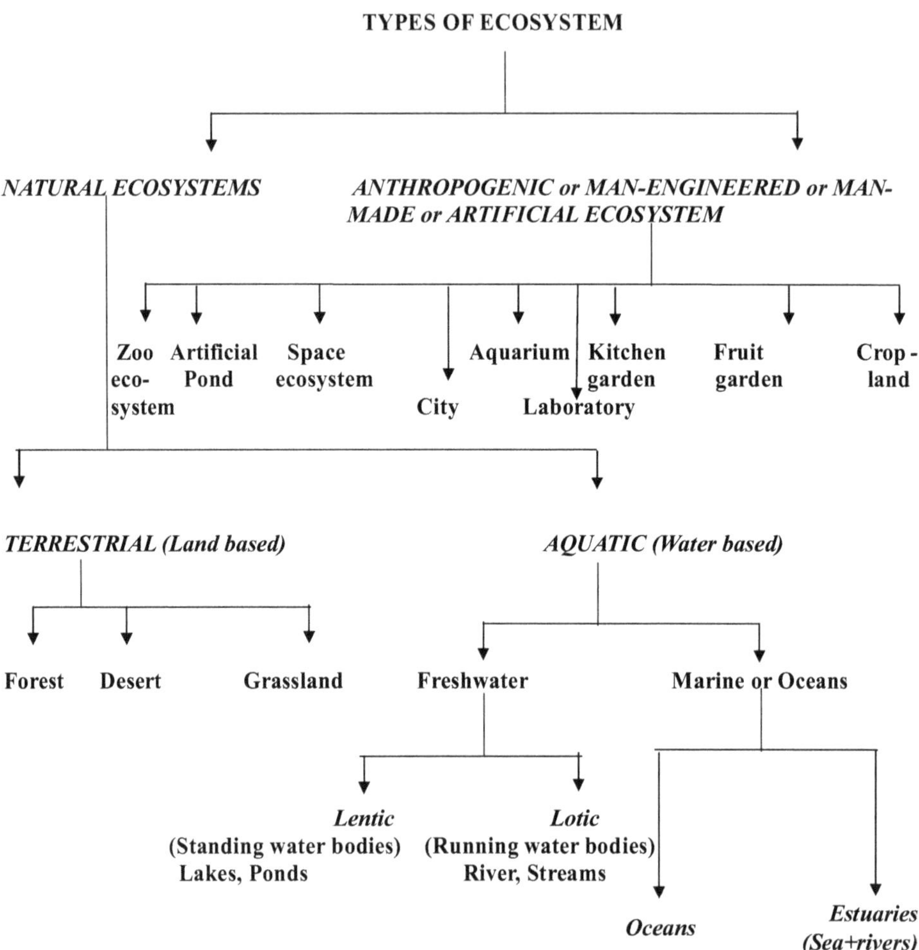

Figure 1.11: Classification of Ecosystems.

Table 1.4: Structure and Function or Ecology of Terrestrial Ecosystems.

Teresstrial Ecosystem	*Abiotic Factors*	*Producers*	*Primary Consumers*	*Secondary Consumers*	*Tertiary Consumers*	*Decomposers*
Forest ecosystem ***(MTU, 2011)***	**Air:** Gases **Water:** Minerals **Sunlight:** Sunlight is the limiting factor for the growth of plants because its penetration is less due to thick canopy of trees. **Temperature:** Varies (Changes) **Soil:** Nutrients (Nitrates+Phosphates), Dead organic matter	**Trees** *e.g.* Neem Ashok, bargad, Arjun, Sal, *Tectona grandis, Shorea robusta, Lagerstroemia parviflora, Thuja, Pinus, Cedrua, Rhododendron etc.* **(Figure 1.12)**	Elephants, Deer, Neelgai, Rabbit, Horse, Giraffe, Flies, Squirre, ants, flies, leafhoppers, bugs, spiders.	Fox, Snake, Lizard, Wolf	Lion, Tiger	**Bacteria** *Pseudomonas* **Fungus** *Actinomycetes* group, *Aspergillus*

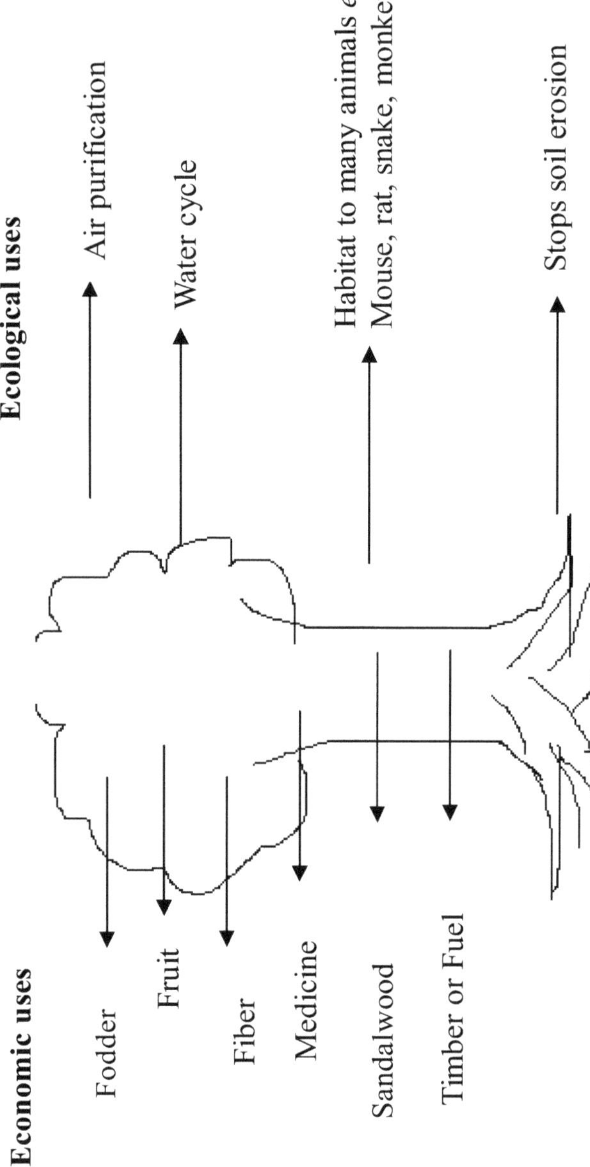

Figure 1.12: Ecosystem Services Provided by Tree in Forest Ecosystem.

Table 1.5: Structure and Ecology of Desert Ecosystem.

Teresstrial Ecosystem	*Abiotic Factors*	*Producers*	*Primary Consumers*	*Secondary Consumers*	*Tertiary Consumers*	*Decomposers*
Desert ecosystem ***(MTU, Dec 2011)***	**Air:** Gases **Rainfall:** very less ($\leq$25) **Sunlight:** enough **Temperature:** High **Soil:** sand, very less water-holding capacity	**Lower plants** lichens, xerophytic mosses **Cactus** **Very less Trees** *e.g.* Babool (*Acacia*), Ber (*Zizhyphus*), Khejri (*Prosopis cineraria*) **Adaptations** Leaves are reduced to spines to reduce evaporation of water. Wax-coating on plant surface **(Figures 1.13 & 1.14).** Roots spread on soil surface (shallow root system to capture rainwater quickly), do not go deep underground.	Nocturnal rodents (adaptation to avoid excessive heat) and Birds, Flies, Camel (ship of desert) Camel has a soft sponge in the foot. Camel can store fat in hump.	Fox, Snake, Lizard, Wolf	Lion, Tiger	**Thermophillic** microorganisms **Bacteria** *Pseudomonas* **Fungus** *Aspergillus*

Table 1.6: Structure and Ecology of Grassland Ecosystem.

Teresstrial Ecosystem	*Abiotic Factors*	*Producers*	*Primary Consumers*	*Secondary Consumers*	*Tertiary Consumers*	*Decomposers*
Grassland ecosystem (Figure 15)	**Air:** Gases **Rainfall Sunlight:** sufficient **Temperature:** moderate **Soil:** Nutrient-rich	**Grasses** *e.g. Cynodon, Saccharum*	Grazing animals Deer, Rabbit, Mouse, Insects, Buffalo	Fox, Snake, Lizard, Jackal	Hawk, Vulture	**Bacteria** *Pseudomonas* **Fungus** Species of *Aspergillus*, Penicillum, Mucor, Rhizopus

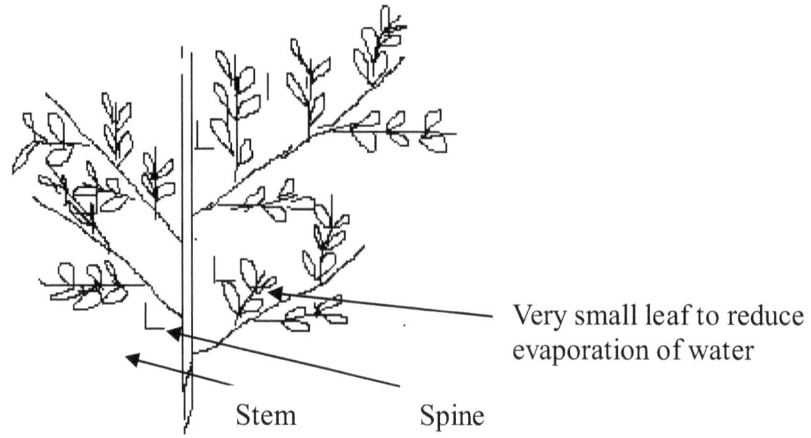

Figure 1.13: Ecological Adaptations by Xerophytic Vegetation (Babool tree or *Acacia* sp.).

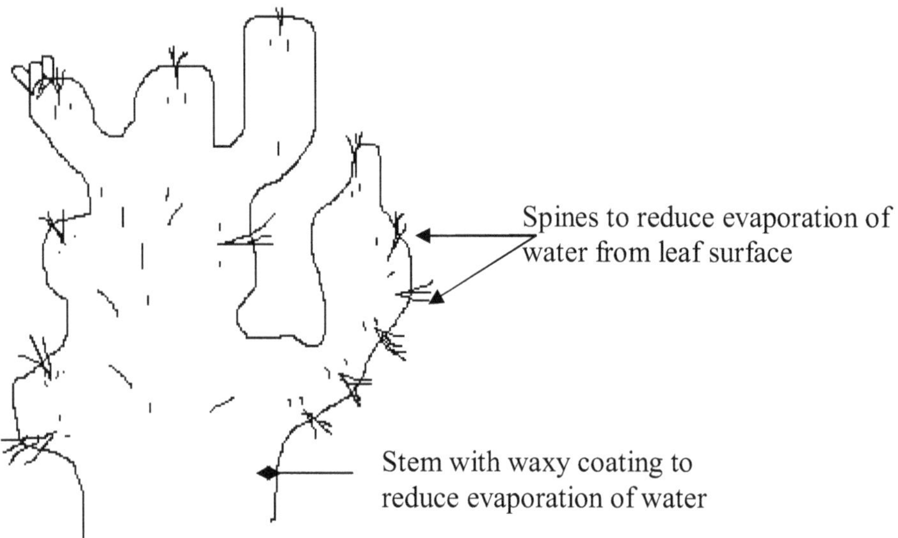

Figure 1.14: Ecological Adaptations by Xerophytic Vegetation (Cactus or *Opuntia* sp.).

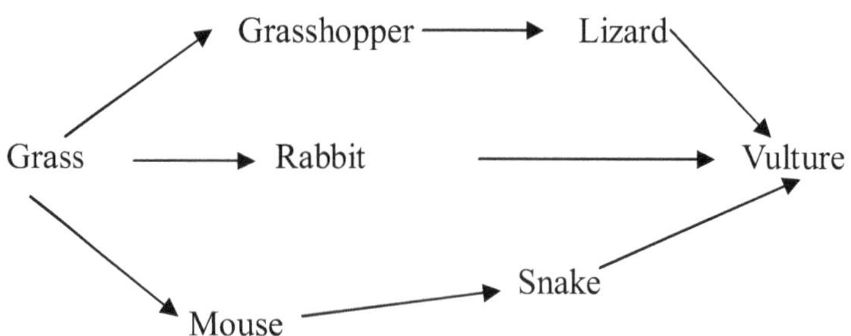

Figure 1.15: Food-web of Grassland Ecosystem.

Table 1.7: Structure of Pond or Lake Ecosystem.

Aquatic Ecosystem	*Abiotic Factors*	*Producers*	*Primary Consumers*	*Secondary Consumers*	*Tertiary Consumers*	*Decomposers*
Pond or Lake (closed ecosystem ***i.e.*** high nutrients are present in the soil). **(Figure 1.18)**	**Sunlight:** Penetration is less to bottom. Its availability is more at the surface, thus known as limnetic zone. **Temperature:** Moderate **Soil:** Nutrient-rich **Water:** Dissolved calcium, nitrogen, phosphate, Organic substances: amino acids, humic acids, dissolved oxygen and CO_2.	Small water plants (phytoplanktons) *e.g.* Algae: *Ulothrix*, *Spiro gyra*, *Cladophera*. Chlorococales: *Cosmorium*, *Pandorina*, Volvox, Microcystis, Spirulina *etc.* Green plants (Macrophytes: Rooted,partly or completely submerged, floating and emergent hydrophytes *e.g.* Lotus (*Nelumbo*), Salvinia, Wolfia, Pistia, Hydrilla, *Vallisneria*, Trapa, *Nymphaea*, *Jussiaea*, *Typha*, *Scirpus*, *Ranunculus* , Hydrilla, Trapa, Lemna).	Small water animal-like organisms (zooplanktons) *e.g.* *Euglena*, *Coleps*, *Brachionus*, Insect larvae.	Insects, small fish, mites, beetles.	Big fish *e.g.* *Game fish*	**Bacteria** *Pseudomonas* **Fungus** (*actinomycetes* is a common bacteria) *Aspergillus*, *Penicillum.*

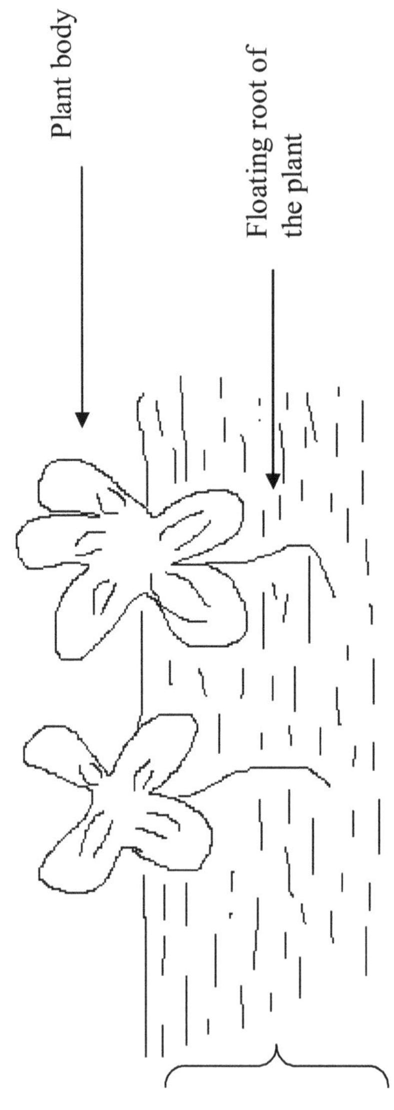

Figure 1.16: Free Floating Small Water Plants or Fronds (Phytoplanktons *e.g. Lemna* sp.).

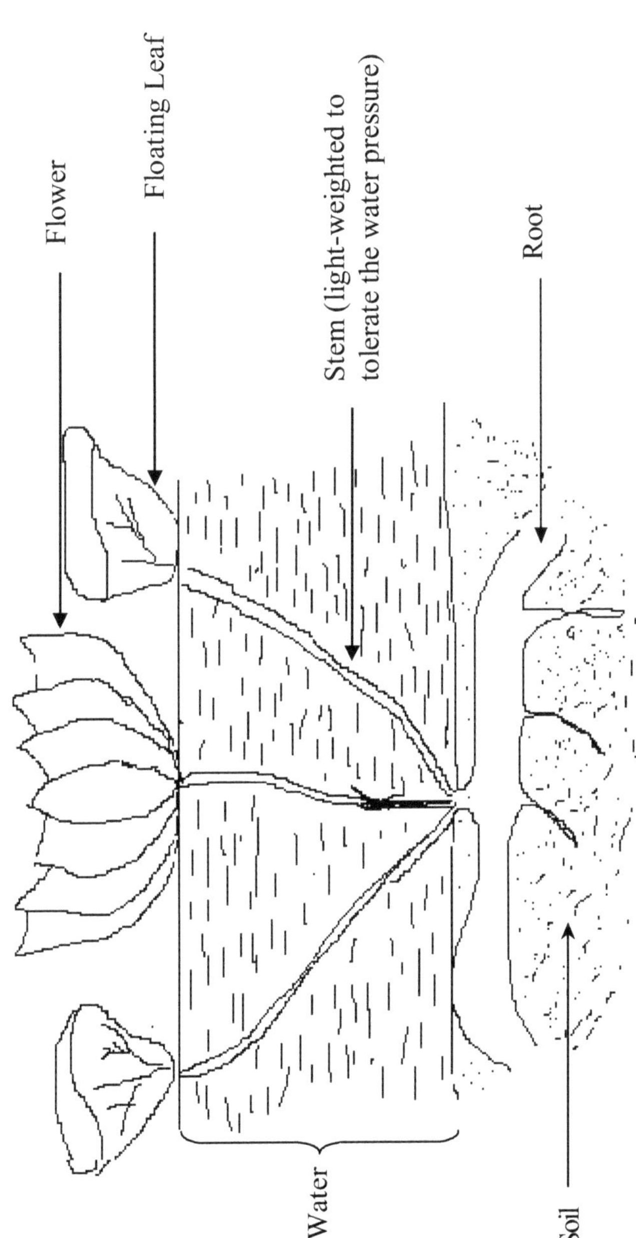

Figure 1.17: Rooted Floating Water Plants (*e.g.* Lotus or *Nelumbo* sp.).

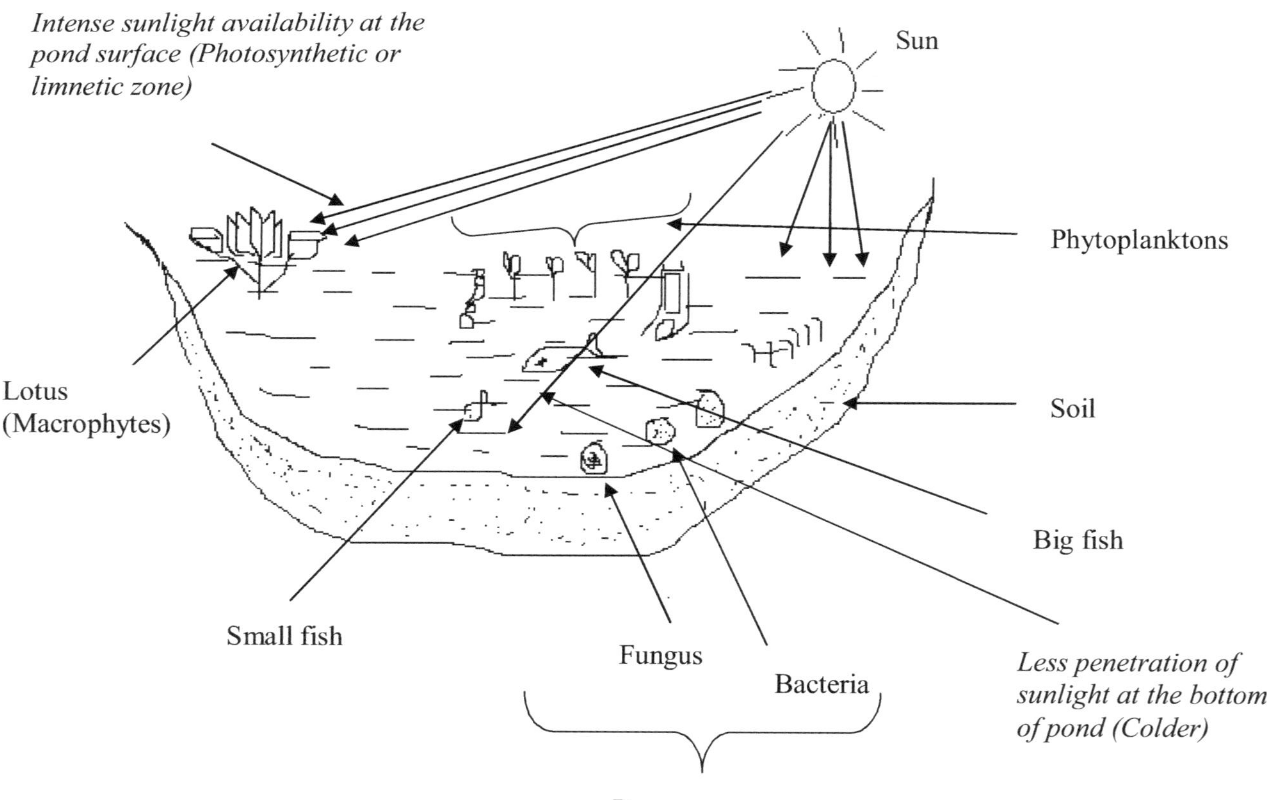

Figure 1.18: Diagramatic Representation of Pond Ecology.

Table 1.8: Structure of River Ecosystem.

Aquatic Ecosystem	*Abiotic Factors*	*Producers*	*Primary Consumers*	*Secondary Consumers*	*Tertiary Consumers*	*Decomposers*
River or Stream (Open ecosystem *i.e.* nutrients keep dissolving from air and soil, when river runs).	**Sunlight:** sufficient. **Temperature:** moderate. **Soil:** Nutrient-rich. **Water:** Well oxygenated.	Small water plants (phytoplanktons) *e.g.* Algae, Green plants, Water grasses (Macro-phytes *e.g.* Lotus, Hydrilla, Trapa, Lemna).	Small water animal like organism (zooplanktons) *e.g.* insect larvae.	Insects, small fish.	Big fish, Crocodile.	**Bacteria** *Pseudomonas.* **Fungus** *Aspergillus, Penicillum.*

Table 1.9: Structure of Marine Ecosystem.

Aquatic Ecosystem	*Abiotic Factors*	*Producers*	*Primary Consumers*	*Secondary Consumers*	*Tertiary Consumers*	*Decomposers*
Marine ecosystem (closed ecosystem)	**Water** High salt Occurrence of **Tides** **Sunlight:** Sufficient at the surface water zone, very less penetration at the bottom of sea *i.e.* colder zone. **Soil:** Nutrient-rich.	Small water plants (phytoplanktons) *e.g.* Algae (Diatoms, Dinoflagellates) Large plants (Macro-phytes) *e.g.* Sea weeds, Sea grasses.	Crustaceans, Molluscs, Fish.	Small fish (carnivorous) *e.g. Herring, Mackerel, Shad.*	Big fish *e.g.* Whale, Dolphin, *Haddock, Cod, Halibut.*	**Bacteria** *Pseudomonas* **Fungus** *Aspergillus, Penicillum.*

Table 1.10: Structure of Estuaries Ecosystem.

Aquatic Ecosystem	*Abiotic Factors*	*Producers*	*Primary Consumers*	*Secondary Consumers*	*Tertiary Consumers*	*Decomposers*
Estuaries (Biodiversity rich ecosystem: Fresh water + Marine water) **or** (Rivers+ Oceans, Tidal marshes, Coastal bays, Water bodies behind barrier beaches)	River's mouth where rivers enter oceans **Water**- Very large amount of nutrients	Small water plants (phytoplanktons) Large plants (Macrophytes) *e.g.* Sea weeds, Sea grasses, marsh grasses	Crustaceans, Molluscs, Fish, Oysters	Crabs, small fish	Big fish	**Bacteria** *Pseudomonas* **Fungus** *Aspergillus, Penicillum*

Roles played by an ecosystem in the biosphere are designated in terms of energy flow, material flow, food-chain and food-web and ecosystem services.

(i) Ecosystem Services

Biogeochemical cycling in the ecosystem helps in many important processes useful for life. *e.g.*

- Air purification
- Pollination
- Water cycle
- Soil development

(ii) Energy Flow in an Ecosystem (MTU, 2013-14)

Energy flow is ***uni-directional or non-cyclic*** and continuous in nature, in an ecosystem (**Figure 1.19**) described in following steps:

1. Sun is the ultimate source of energy in an ecosystem.
2. Producers (green plants) carry out photosynthesis to make their own food in the presence of CO_2, water and sunlight.
3. Energy is transferred from producers to herbivores to carnivores and to decomposers.
4. Loss of energy occurs during its transfer from one organism to another.

(iii) Material (Nutrient or Mineral) Flow in an Ecosystem or Natural Cycle or Biogeochemical Cycling

Flow of nutrients is bidirectional or cyclic in nature. Minerals are transferred from air, water and soil to producers to herbivores to carnivores to decomposers and finally to air, water and soil (**Figure 1.20**).

(iv) Productivity

The rate at which organic matter or biomass is produced is known as productivity. (a) **Primary productivity** is the rate at which green plants carry out photosynthesis to produce organic matter, in the presence of CO_2, water and sunlight. It is measured in g m^{-2} for dry matter production and kcalm^{-2}year^{-1} for energy production. The primary productivity is classified as Gross Primary Productivity (GPP) and Net Primary Productivity (NPP). (b) **Gross primary productivity** is the rate at which green plants capture total energy or produce total organic matter. (c) **Net Primary Productivity**: is the rate at which green plants store organic matter after respiration and maintenance per unit area and time. NPP of tropical rain forests is very high as compared to deserts.

Net Primary Productivity =

Gross primary productivity – Loss due to respiration and maintenance

The rate at which biomass of consumers increases per unit area and time is known as secondary productivity.

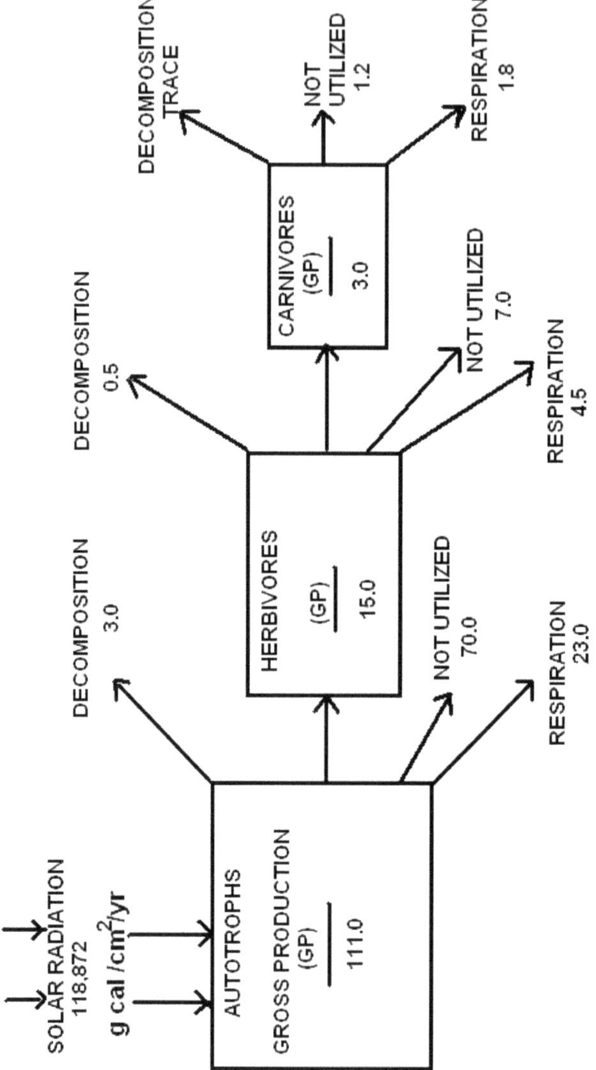

ENERGY FLOW DIAGRAM FOR A LAKE IN [g cal /cm^2/yr]

Figure 1.19: Energy Flow is Unidirectional *i.e.* from Source of Energy (Sun), it Flows to Autotrophs, Herbivores, Carnivores and Decomposers (MTU, 2013-14).

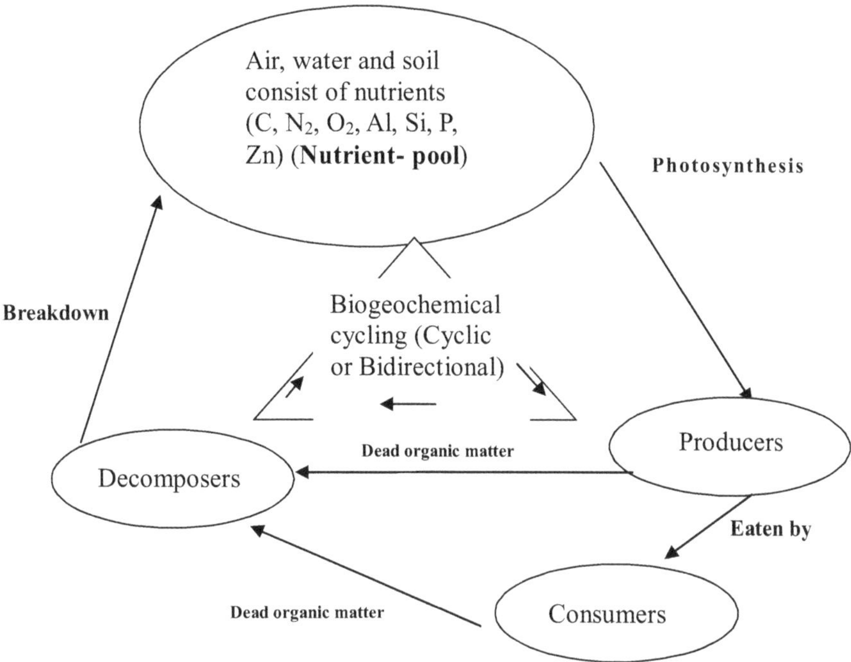

Figure 1.20: Biogeochemical Cycling in an Ecosystem.

(v) Ecosystem Maintains *Balance* in Nature by Food-Web

It maintains *balance* or stability in an ecosystem. In ecosystems, a single food-chain cannot exist in isolation, neither the single grazing food-chains nor the detritus food-chain can exist. The decomposer and grazing food-chains are interlinked with each other because the waste from grazing food chain enters into decomposer food chain. Nutrients are recycled by decomposer food-chain or are reincorporated into the first trophic level or base of the grazing food-chain.

1.4.7. Biogeochemical Cycling/Nutrient Cycling *(UPTU, June 2010)*

Nutrients or minerals are transferred from air, water and soil to producers to herbivores to carnivores to decomposers and finally to air, water and soil. Thus nutrient flow is bidirectional or cyclic and continuous in nature.

1.4.7.1 Sulphur Cycle

Sulphur (S) forms linkages between the polypeptide chains which combine together to form protein. It helps in protein synthesis. Sulphur exists in several Oxidation states in nature: H_2S, SO_2^-, SO_4^{-2}. Sulphur exists in elemental form also. Important steps of sulphur cycle (**Figure 1.21**) are as follows:

- ✰ Sulphur is also locked in coal and petroleum and released as SO_2 when burned.

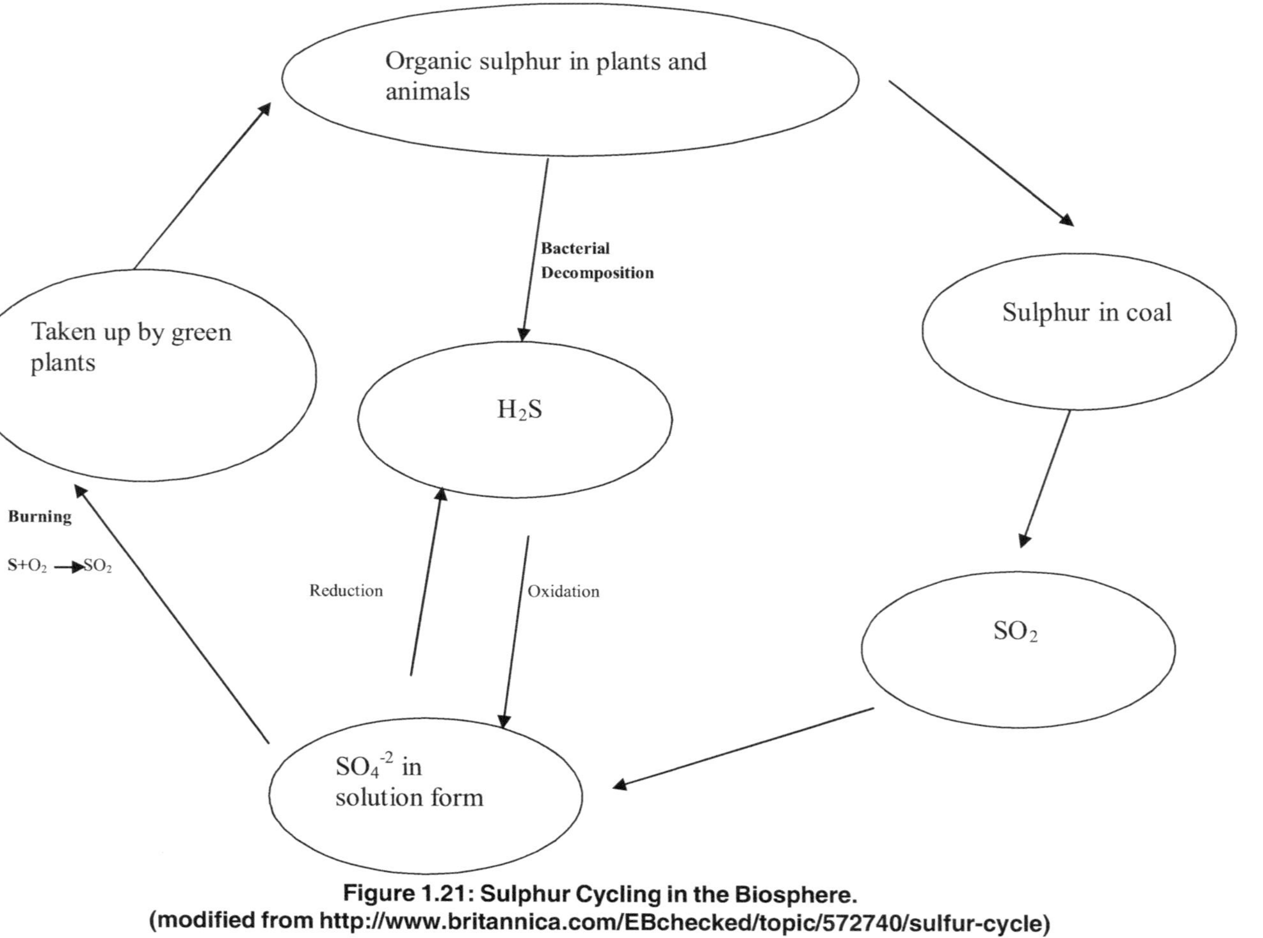

Figure 1.21: Sulphur Cycling in the Biosphere.
(modified from http://www.britannica.com/EBchecked/topic/572740/sulfur-cycle)

☆ Suphite ions convert into sulphate ions in solution form. Plants absorb sulphates from soil and water.

☆ Bacteria decompose the dead and decaying parts of plants and animals. H_2S is released in the bacterial decomposition.

☆ H_2S is converted to SO_4 by bacteria of genus *Thiobacillus*, which is absorbed by plants (nutrient).

1.4.7.2 Carbon Cycle

Carbon forms the major components of the organic matter *i.e.* plant, animal body **(Figure 22).**

1. CO_2 is present in earth's atmosphere (0.035 per cent).
2. Green plants carry out photosynthesis in the presence of CO_2, water and sunlight. They make their own food.
3. Animals eat green plants.
4. During respiration, plants and animals return CO_2 to atmosphere.
5. Burning of fossil fuels in industry releases CO_2 to atmosphere.

So the carbon-cycle is bidirectional or cyclic.

1.4.7.3 Nitrogen Cycle (*UPTU, 2010, 2011*)

Nitrogen is the essential macro-nutrient (required in large amount) for the plant and animal growth. The major amount of nitrogen is gaseous and thus present in atmosphere. Pants cannot absorb the gas. Thus it is converted to some other absorbable form by nature described below: **(Figure 1.23)**

Processes in Nitrogen Cycle

(1) Nitrogen Fixation

The process of converting atmospheric nitrogen to the form which plants can absorb is known Nitrogen Fixation.

Types of Nitrogen fixation

☆ **Atmospheric N-Fixation**

$$\text{Nitrogen molecules + Oxygen molecules} \xrightarrow{\text{(By lightning).}} \text{Nitrates (taken by plants)}$$

☆ **Biological/Biotic N–Fixation**

By certain micro-organisms:

(i) ***Rhizobium*** is a symbiotic bacterium (Rhizobium bacteria lives in the root nodules of leguminous plants).

(ii) ***Azotobacter*** and ***Clostridium*** are free living bacteria.

(iii) ***Anabaena*** and ***Nostoc*** are some cyanobacteria

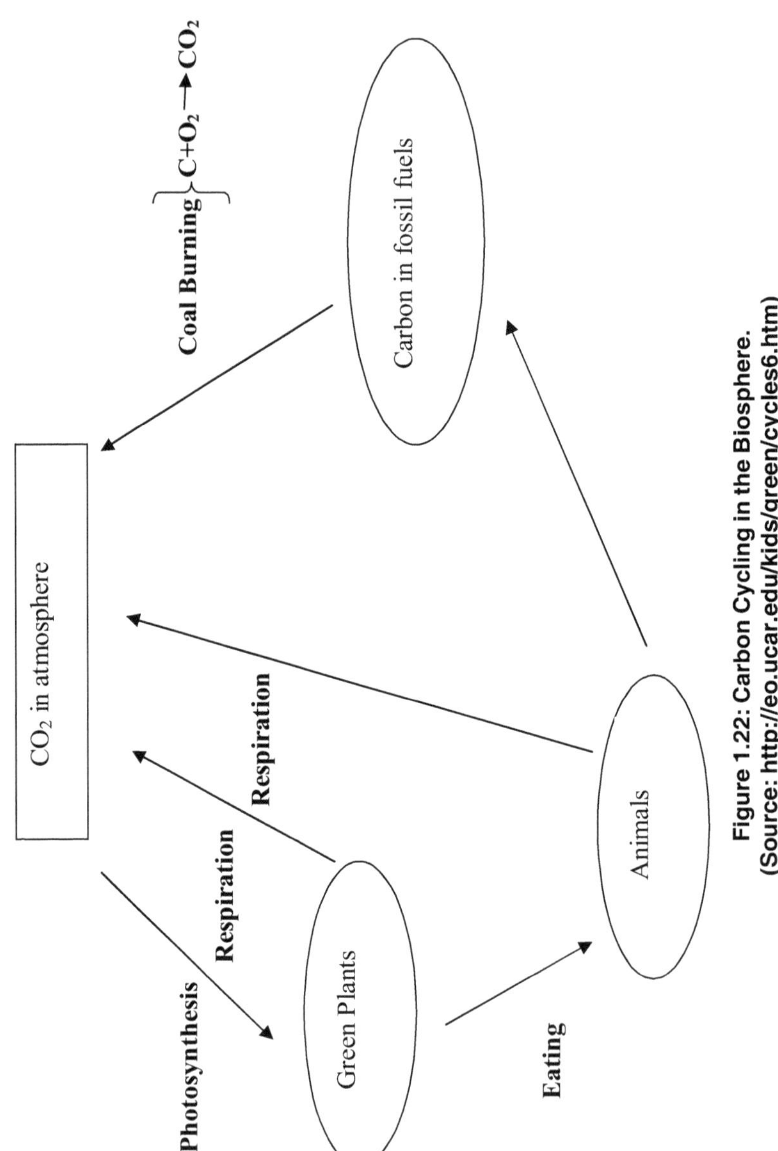

Figure 1.22: Carbon Cycling in the Biosphere.
(Source: http://eo.ucar.edu/kids/green/cycles6.htm)

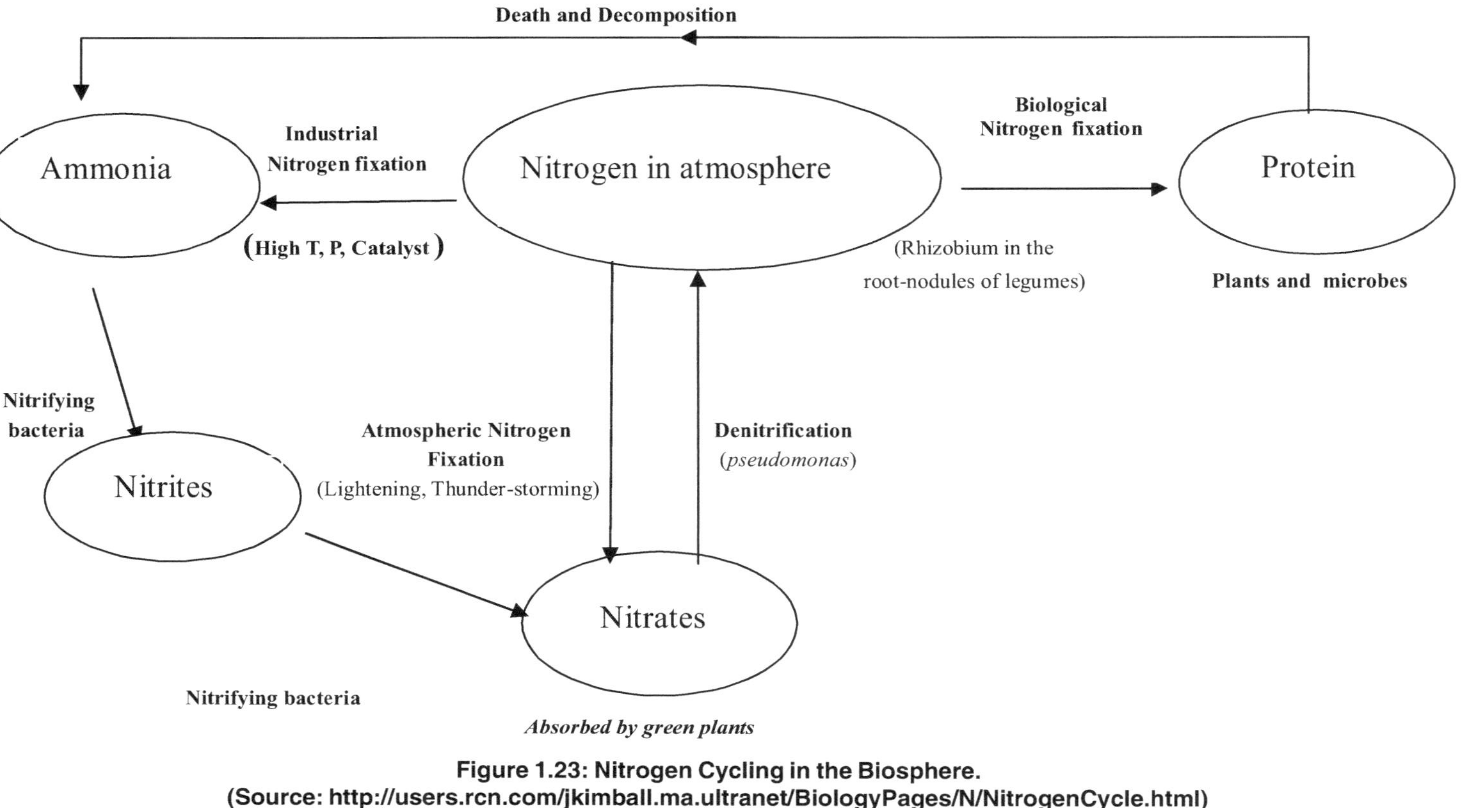

Figure 1.23: Nitrogen Cycling in the Biosphere.
(Source: http://users.rcn.com/jkimball.ma.ultranet/BiologyPages/N/NitrogenCycle.html)

☆ **Industrial N–Fixation**

In industries, at high temperature, pressure and catalyst.

$$\text{Nitrogen} + \text{Hydrogen} \xrightarrow[\text{Catalyst}]{\text{High T, P}} \text{Ammonia}$$

(2) Decomposition

Decomposers breakdown the dead and decaying organic matter. The most stable product of decomposition is ammonia.

(3) Nitrification

Nitrifying bacteria convert ammonia to nitrite (*Nitrosomonas*) to nitrate (*Nitrobactor*).

(4) Denitrification

Denitrifying bacteria returns Nitrogen to atmosphere (in the absence of oxygen). *e.g. Pseudomonas*

1.4.8 The Y-Shaped Energy Flow Model

E.P. Odum (1983) introduced Y-shaped or Two-channel energy flow model designed in such a way that it shows importance of decomposers in the process of energy flow. It is applicable to both aquatic and terrestrial ecosystems. This model also based upon the basic trophic structure of any ecosystem. It links grazing and detritus food chain in time and space (**Figure 1.24**).

Physical Significance of Energy Flow

In nature, food-chains are interrelated with each other very closely. The complexity of food chains forms food-web. Thus multi-channel energy flow runs in the biosphere, which makes it difficult to be measured.

1.4.9 Ecological Pyramids (MTU, 2013-14)

Pictorial representation of transfer of energy or number of individuals or dry weight of individuals from one trophic level to another, in an ecosystem is known as ecological pyramids. The term ecological pyramid is given by Elton Charles (1927), thus ecological pyramids are also known as eltonion pyramids.

Types of Ecological Pyramids

(i) Pyramid of Numbers

Pictorial representation of transfer of number of individuals from one trophic level to another, per unit area, in an ecosystem is known as ecological pyramids of number.

Grassland Ecosystem

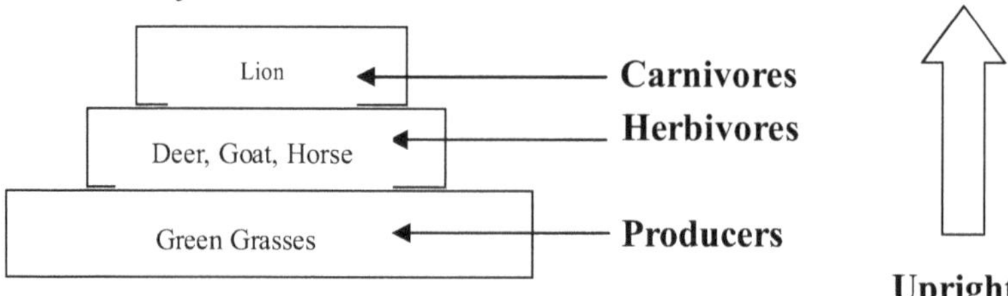

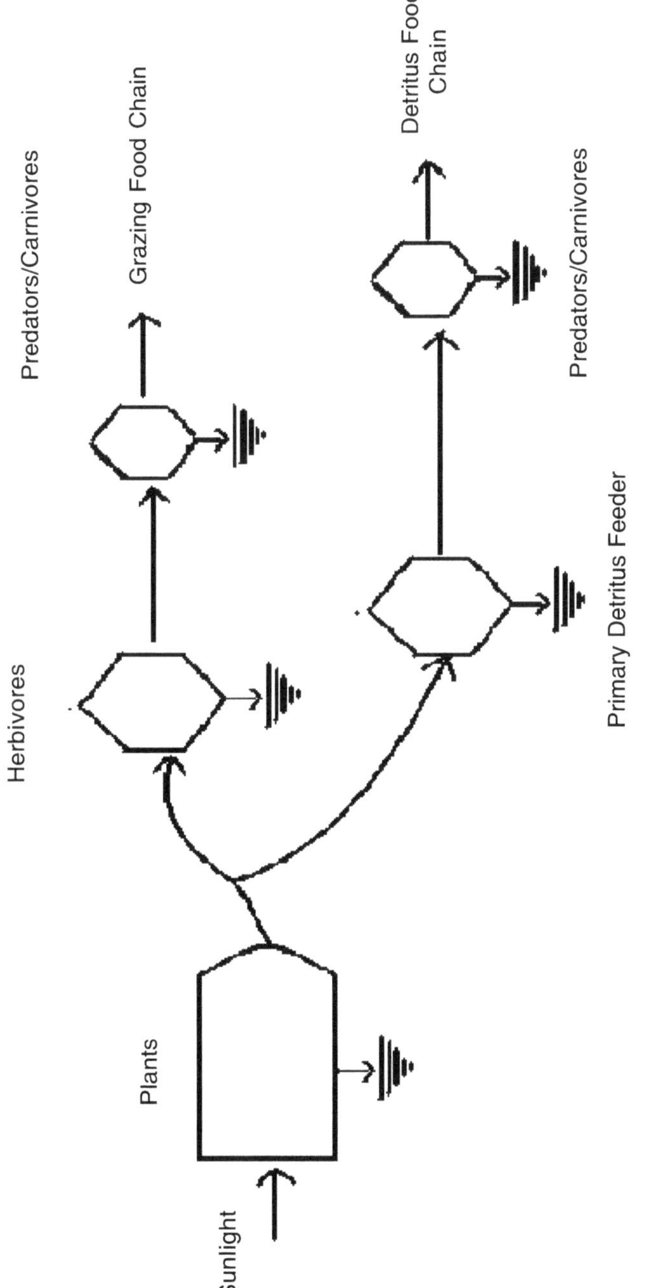

Figure 1.26: Y-shaped Energy Flow Model.

Pond Ecosystem

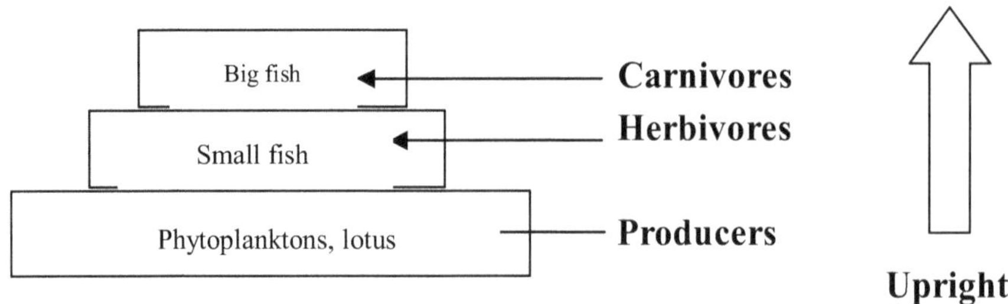

Forest Ecosystem

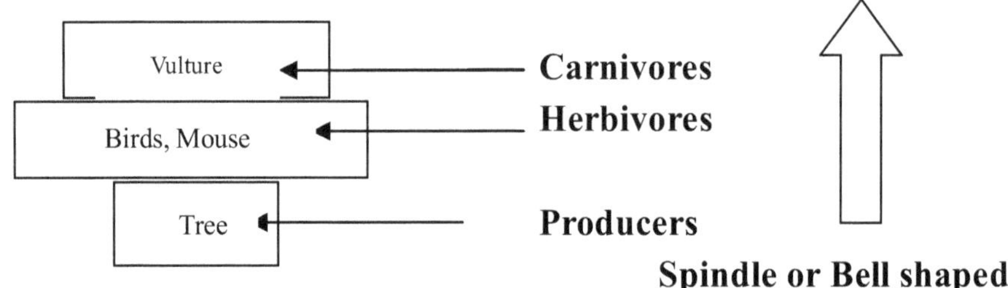

Tree Ecosystem

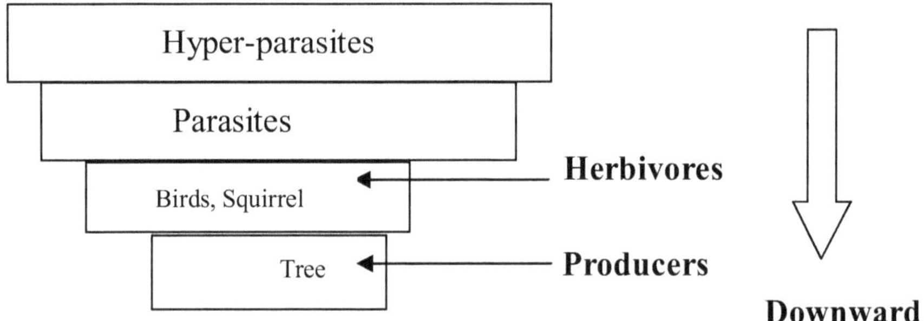

(ii) Pyramid of Biomass

Pictorial representation of transfer of dryweight of individuals from one trophic level to another, in an ecosystem is known as ecological pyramids.

Grassland Ecosystem

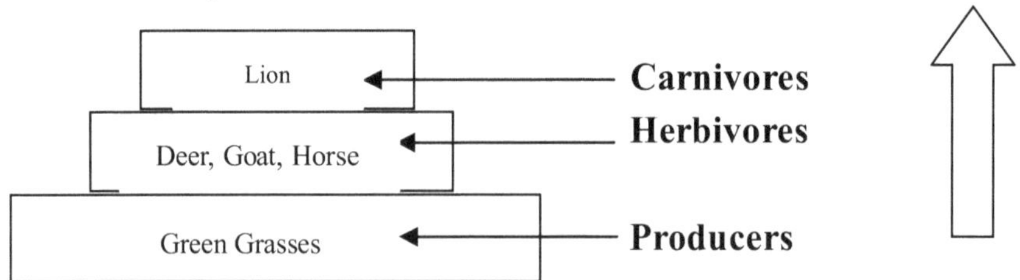

Pond Ecosystem

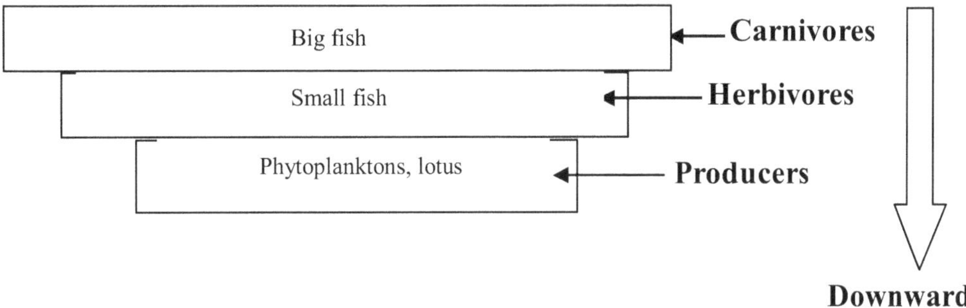

(iii) Pyramid of Energy (MTU, 2010, 2011)

Pictorial representation of transfer of energy from one trophic level to another, in an ecosystem is known as ecological pyramids. Pyramid of energy is always upright because loss of energy occurs during its transfer from one trophic level to another into respiration, decomposition and heat losses in an ecosystem.

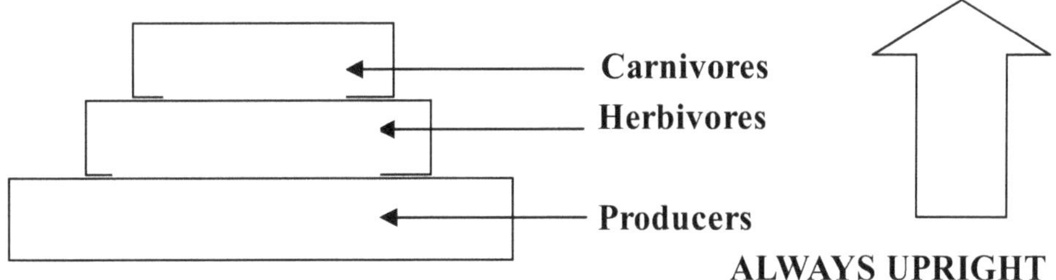

Figure 1.25: Ecological pyramids: Pyramid of number: in pond and grassland, forest ecosystem and parasitic food chain, Pyramid of biomass in pond and grassland:, Pyramid of energy.

1.4.10 Ecorestoration of Damaged Ecosystems

The branch of ecology which enhance the understanding of natural ecosystems and ways to protect and restore them when damaged or degraded due to human interference is known as ecorestoration or restoration ecology or conservation biology.

Restoring natural ecosystems that have been damaged due to anthropogenic interference is covered under the integrated principles of sustainability *i.e.* ecorestoration is a tool for sustainable development. We cannot afford to exploit the land, deplete the natural ecosystems and resources due to severe natural resource crisis. Services and goods provided by ecosystems fulfill the growing demands of population increase. To support the needs of all the individuals of present generation and the future generation we must ensure wasteland re-claimation, replant forests, rebuild wetlands, and restore grassland and pastures. Thus, ecorestoration helps in sustaining the biosphere.

1.4.10.1 Concept of Ecorestoration

The process of ecological succession forms the basis of ecorestoration. Ecosystems mature on barren or relatively lifeless substrate by the natural process known as natural ecological succession. Natural ecological succession is of two types: **Primary succession** and **secondary succession**. Primary succession is the process wherein mature ecosystems are formed on barren substrate in which no life previously existed. For-example lichens (fungus and algae or fungus and photosynthetic bacteria) grow on barren rock, known as **pioneer community**, with the following steps of ecological succession:

(i) Lichens adhere to the rock surface, absorbing moisture from rain and carbon-dioxide from air.

(ii) They do photosynthesis to produce organic nutrients.

(iii) They secrete a weak acid, carbonic acid which dissolves rock.

(iv) Tiny insects and lichens invade the rock.

(v) Their dead and decaying parts form the soil.

(vi) Mosses get established to form roots known as **intermediate community**.

(vii) As soil becomes more fertile herbs and small shrubs grow.

(viii) Eventually, the final stable stage of a stable ecosystem is the forest (end-point of natural succession).

1.4.10.2 Secondary Succession Forms the Basis of Natural Ecosystem Restoration

The nature's way of restoring the ecosystem *i.e.* bringing to its natural vegetative stage by sequential development of biotic communities (*i.e.* before the complete or partial destruction of an existing community) is secondary succession. It occurs faster than primary succession (but still takes a very long passage of time *i.e.* more than 70 years). Ecorestoration of degraded farm-land is described in following steps:

(i) Very small grass colonizes in the first year.

(ii) Tall grass, herbs and small plants establish in 1-3 years.

(iii) Small pine trees invade the restoring farmland during 3-10 years.

(iv) Pine forest is established in 10-30 years.

(v) Hardwoods *e.g. Oak, Neem, Peeple, Bargad* invade 30- 70 years.

(vi) Trees establish as dominant species in more than 70 years of duration and forms stable species known as **climax** stage because they are capable of restoring the ecosystem damage (minimize soil erosion, maintain air and water cycle, provide habitat to many species, shade to small plants, helps in recycling of nutrients *etc.*).

1.4.10.3 Ecorestoration Technology

Environmental researchers and scientists have opted for bioremediation studies to facilitate the nature's restoration process of degraded ecosystems. The process of removal of toxic substances from air, water and soil environment with the help of living organisms is known as bioremediation. *E.g.* air pollution tolerant trees (Neem, Peepal, Bargad) detoxify the polluted air; water hyacinth absorbs heavy metals from water, hedge plants absorb noise waves; *Bauhinia variegate* (kachnar), morus alba, peepal absorbs heavy metals from soil.

1.4.10.4 Bioremediation as a Viable Tool for Ecosystem Restoration

As discussed, nature's eco-restoration process takes a long passage of time in reclaiming the degraded ecosystem. Understanding the mechanism of eco-restoration and facilitating the process of ecosystem recovery is carried out under the bioremediation studies for restoration ecology. Bioremediation is classified into the following categories, according to the biological agents used for detoxification of the environment:

(i) Phytoremediation

(ii) Microbial remediation

(iii) Zooremediation

1.4.10.4.1 Phytoremediation

The Greek word "*phyto*" means plant and the Latin word "*remedium*" means to cure or restore. Phytoremediation is defined as "the process of restoration of quality of environment by the application of plants." *i.e.* use of green plants based systems to detoxify or remove toxic, substances from contaminated air, water, soil and sediments, also called as **green clean**.

Types of Phytoremediation

Following forms of phytoremediation are applicable in different technologies:

(i) Phytovolatilization

Certain transgenic plants reduce hazardous ionic and methylated forms of mercury to elemental mercury, which is then volatilized. The technique utilized is known as phytovolatilization. Selenium phytoremediation is also done by phytovolatilization.

(ii) Phytostabilization

In this form of phytoremediation, agronomic techniques are used to stabilize contaminated sites (*in-situ*) *i.e. checking the spread of contaminant in the different segments of environment*, with following steps:

(a) Soil amendments are applied to contaminated soil to reduce the bioavailability of the contaminants.

(b) Vegetation is grown on the site to reduce the movement of contaminants with water-runoff and soil erosion.

(c) Plants are grown to maximize the root uptake of contaminants.

(iii) Phytoextraction

It is the most important form of phytoremediation in which toxic and non-biodegradable (persistant) contaminants *e.g. Heavy metals* are extracted from the environment by growing green plants on the contaminated soil.

The toxic compounds are captured by the special compounds present in the plants, by the mechanism of chelating known as **phytochelatins**.Plants were shown to exhibit metal binding proteins known as **metallothioneins**. Researchers have found that heavy metal tolerant plants exhibit higher amounts of proteins to accumulate higher concentrations of metals as compared to other plants.

1.4.10.4.2 Restoration of Wasteland by Phytoremediation

Due to industrialization, modern agriculture and research activities, lot of toxic solid waste is disposed on the land area, causing land pollution. The plants exhibiting phytoremediation should have following characteristics:

(a) They should be fast growing plants.

(b) They should have high biomass potential to accumulate large amount of toxicants. They normally show a property of binding the metal into their root tissues, transporting upward from their roots.

(c) They should be locally growing tress to be able to adapt to the climatic conditions.

The potential phytoremediators are given in Table.

Table: Potential Phytoremediators

Potential Phytoremediators	*Ecorestoration*	*Remarks*
Brassicaceae: The family of scavengers *i.e.* Members of Brassica plant family (*Brassica juncea; 'sarson'*)	Adapted to absorb toxic heavy metals to levels 30 to 1000 times higher than concentration in the surrounding soil, through their roots.	Antimony, Arsenic, Barium, Beryllium, Cadmium, Cerium, Cesium, Chromium, Cobalt, Copper, Gold, Indium, Lead, Manganese, Mercury, Molybdenum, Nickel, Palladium, Plutonium, Rubidium, Ruthenium, Selenium, Silver, Strontium, Thallium, Tin, Uranium, Vanadium,Yttrium, Zinc.
Sebertia acuminate: The Nickel Lover (tree)	Hyper-accumulate Nickel 20 per cent of its dry weight	When slashed, it bleeds a jade-green liquid containing nickel.
Pine tree	Absorbs Beryllium.	Bark of the tree absorbs 100 times more metal than soil.

(a) Decontamination of Heavy Metals

Phytoremediation of Lead (Pb), Cadmium (Cd), Arsenic (As), Mercury (Hg), Chromium (Cr), Nickel (Ni) *etc.* is an important application. The plants have been shown to exhibit phytoremediation potential in Table.

(b) Phytomining

Extraction of metals of economic importance *i.e.* precious metals (Ni, Cu and others) with the help of phytoremediators, is known as phytomining.

1.4.10.4.3 Restoration of Aquatic Ecosystems by Phytoremediation

Among floating macrophytes, water hyacinth and salvinia are fast growing weeds. Their higher productivity and resilience *i.e.* ability to accumulate very large amount of toxic substances (viz. heavy metals) make them ideal macrophytes for wastewater treatment (secondary and tertiary waste water treatment). The biomass is harvested frequently to assure maximum productivity and removal of toxic substances. Macrophytes remove contaminants from waste water through:

(i) Roots and Stems in Water Column

They provide surface area for bacterial growth and media for filtration and absorption of solids. Natural polymers are released to facilitate flocculation sedimentation.

(ii) Stems and Leaves at or above the Water Surface

They facilitate the transfer of gases and heat between atmosphere and water. Some plants are very efficient oxygen pumps. They also reduce the effects of wind on water and regulate the penetration of sunlight and thus curtail the growth of algae.

1.4.10.4.3.1 Rhizofiltration

Removal of toxic compounds in the contaminated water environment, by the filtration process of the roots of the plants is known as rhizofiltration. Roots of the plants act as biofilters *i.e.* roots help absorb toxic compound or metals from contaminated water and accumulate them. Contaminants are removed from the plants by harvesting (collecting from) the root biomass.

Potential Phytoremediators of Aquatic Ecosystems	*Detoxification Potential*	*Remarks*
Sunflower	The uranium scavenger and other radioactive elements	Sunflowers soak uranium, purifying radioactive water, near radioactive sites. In a research (for 10 days), ability of sunflower to soak up 95 per cent radioactive elements (Strontium and Cesium) at Chernobyl was found.
Floating aquatic plants Water hyacinth (*Eichhornia sp.*)		Extensive root system has excellent filtration and bacterial support potential. cannot survive in harsh winters (near 0°C).
Water primrose (*Ludwigia sp.*)		
Acer platanoides, *Chrysanthemum* sp.		
Cassia siamea, Zizyphus mauritiana		Lead enters the plants through stomata. Higher quantity of lead is absorbed when grown near roadside.

1.4.10.5 Restoration of Aquatic Ecosystems: Zooremediation

Zooremediation is the process of decontamination of polluted environment by using animals as bioagents. Animals so far used for bioremediation purposes are different arthropods, fish and other filter feeders in aquatic systems and earthworms in solid organic waste management systems. Use of animals in bioremediation is not very encouraging, except earthworms because there are so many limitations with them *e.g.* fish and arthropods may bioaccumulate the toxic compounds and metals which may be biomagnify in the food-chain creating many problems to the environment.

Arthropods

Daphnia pulex, D. magma, D.similis, Brine shrimp (Artemia salina) etc.

Fish

Hypopthalmichtys molitrix, Aristichtys nobilis, Notemigonus crysoleucas, Tincavulgaris, Pimphales promelas, Mugil *sp. etc.*

1.4.10.6 Restoration of Marine Ecosystems: Microbial Remediation

Microorganisms play a very important role in removing pollutants from the environment. Indian born American microbiologist **Dr. Ananda M. Chakrabarty** developed (patented) a group of genetically engineered two strains of *Pseudomonas*, known as **wonderbugs or superbugs**, which have ability to degrade all four group of compounds present in oil spills (oil eating bacteria).

1.4.10.4.7 Vermiculture and Vermicomposting

Vermicomposting is composting aided by earthworms.

$$\text{Organic waste} \xrightarrow[\textbf{earthworms}]{\textbf{breakdown}} \textbf{simple organic matter}$$

(Nutrient rich **vermicompost**)

Different processes, mechanisms and techniques of bioremediation discussed so far have shown their wide applicability in field. Some of these still need practical acceptability. This emerging technology-bioremediation involves bioagents, which have capability of detoxifying the environment.

Participatory Learning

(a) Objective Questions

1. The study of ecosystem is known as ______.
2. The term ecosystem is given by ______.
3. The term ecology is given by ______.
4. Green plants are ______ of ecosystem.
5. Herbivores are ______ consumers.
6. Decomposers are known as ______.
7. ______ is the ultimate source of energy in an ecosystem.
8. The inter-relationship between living and non-living factors forms a ______.

9. The study of interrelationship between the living and non-living factors is known as ______.

10. Edaphic factor is related with ______. (***UPTU, 2011-12***)

11. The ecological factors related with soil and substratums are called ______ factors.

12. ______ teach us how to treat nature.

13. An ecosystem consists of
    (a) Green plants and animals
    (b) Producers and Consumers
    (c) Green Plants, Animals, decomposers and abiotic environment.
    (d) Green Plants and Decomposers

14. Which one is correct food-chain?
    (a) Phytoplanktons $\rightarrow$ Zooplanktons $\rightarrow$ Fish
    (b) Zooplanktons $\rightarrow$ Phytoplanktons $\rightarrow$ Fish
    (c) Grass $\rightarrow$ Fish $\rightarrow$ Zooplanktons
    (d) None of these

15. Which one of the following is abiotic component of ecosystem?
    (a) Bacteria (b) Plants (c) Humus (d) Fungi

16. Every food-chain starts with a ______.

17. Dead plant parts and animal remains are called ______.

18. Autotrophic planktons are called ______.

19. The pyramid of ______ is always upright.

20. The term ecological pyramid was given by ______.

21. ______ break-down organic matter.

22. The short-term properties of atmosphere at a given place and time is called as ______.

23. The long-term properties of atmosphere at a given place and time is called as ______.

24. The single term by which can define all the physical factors of ecosystem is ______.

25. Termites and worms are the examples of ______.

26. The food chain in which micro-organisms breakdown dead producers is called______.

27. Energy flow is ______ in nature.

28. Material flow is ______in nature in an ecosystem.

True/False

29. Biosphere is composed of atmosphere, lithosphere and hydrosphere.

30. Temperature of thermosphere goes upto 1200°C.

(b) Very Short Answer Type Questions

1. What are the two components or factors or entities of nature or environment?

2. Define ecosystem, ecology, homeostasis, trophic levels, ecological pyramids, ecological succession, environment, environmental sciences, detritus and detrivores.

3. What are the two types of categories of environment?

4. What is the primary or ultimate energy source in an ecosystem?

5. Define nitrogen fixation. Give its types.

6. What are biogenetic nutrients? Also explain mineral cycling or nutrient cycling?

7. What is nutrient pool and cycling pool?

8. Comment "vegetarian diet is universally accepted best diet.".

(c) Short Answer Type Questions

1. Define environment developing relation with human being.

2. Differentiate between:

   (i) Food-chain and food-web
   (ii) Grazing food-chain and Detritus food-chain
   (iii) Climate and weather
   (iv) Climate and microclimate
   (v) Gross Primary Productivity and Net Primary Productivity

3. Explain the hydrological cycle. (***MTU, 2011***)

4. Which ecological pyramids are never inverted? Why? Which type of diet is preferred and why? (***MTU, 2012***)

5. Explain the trophic structure of an ecosystem. What is the physical significance of food-web?

6. Describe the role of rhizobium in nitrogen fixation.

7. Differentiate between nitrifying bacteria and denitrifying bacteria.

8. What is the ultimate energy source in an ecosystem? Or who provides energy to the living organisms?

9. What is the role of decomposers in an ecosystem? Give their alternate names and reasons for their names.

10. What is the importance or role of producers in an ecosystem? Give their alternate names and reasons for their names.

11. Distinguish between the following:
    (i) Biotic and abiotic components of ecosystem
    (ii) Marine ecosystem and estuaries
    (iii) Terrestrial and aquatic ecosystem
    (iv) Natural and man-made ecosystems
    (v) Desert and pond ecosystem
    (vi) Phytoplanktons and zooplanktons
    (vii) Macrophytes and phytoplanktons

12. Why an environmental science is mandatory in engineering curriculum?

13. Discuss the multidisciplinary nature of environmental sciences. (***UPTU, 2008***)

14. Give stratification of lake on the basis of penetration of sunlight. or Differentiate between limnetic zone and littoral zone of lake.

15. Describe the four segments of environment. (***UPTU, 2009-10***)

(d) Long Answer Type Questions

1. Define ecosystem. Explain the types of aquatic ecosystem and terrestrial (desert, grassland, ***UPTU, 2010***) ecosystem. (***UPTU, 2005, MTU 2011, MTU Carryover, 2012-13***)

2. Explain the structure and function of any of the freshwater ecosystems and marine water ecosystems.

3. Explain the role of the biotic components of an ecosystem.

4. Comment "energy flow is unidirectional, non-cyclic and continuous in nature." (***UPTU, 2006, 2008, 2010***) (***MTU, 2013-14***)

5. Define biogeochemical cycling. Explain the different natural cycles in an ecosystem. (***MTU, 2012***).
6. Explain the process of ecological succession or development of a stable ecosystem from a bare land.
7. Explain the concept of balanced ecosystem. How it is maintained? How the mechanism of ecosystem regulation takes place? (***UPTU, 2009***)
8. Describe an ecosystem model. (***MTU, Carryover 2013***)
9. Give the salient features of pond ecosystem, lake ecosystem, marine ecosystem, estuaries, desert, grassland ecosystem.

Answers to Objective Question

1. Ecology	2. AG. Tansley	3. E. Haekal
4. Producers	5. Primary	6. Saprotrophs or detrivores
7. Sun	8. Ecosystem	9. Ecology
10. Soil	11. Edaphic	12. Environmental ethics
13. (c)	14. (a)	15. (c)
16. Producers	17. Detritus	18. Phytoplanktons
19. Energy	20. Charles Elton	21. Decomposers
22. Weather	23. Climate	24. Climate
25. Decomposers	26. Detritus food chain	27. Unidirectional
28. Bidirectional	29. True	30. True

Chapter 2

Effect of Human Activities on Environment

2.1 Food, Shelter, Economic and Social Security

Availability of healthy food, a secured place of residence, economic self dependence and healthy, interactive and cooperative social life are the important factors to maintain healthy environment. This approaching food, shelter, economic and environmental security is an integrated approach to eradicate poverty or achieve equitable distribution of resources or sustainable development.

The poverty line is defined as the amount of income required to satisfy the basic needs of food (also includes sanitation and health care), shelter, clothing and surrounding social environment (also includes education).There are more hungry people in the world today than ever before in human history, and their numbers are growing.

The number of people living in slums and shanty towns is rising, not falling. A growing number lack access to clean water and sanitation and hence are prey to the diseases that arise from this lack. There is some progress, impressive in places. But, on balance, poverty persists and its victims multiply.

2.1.1 Food Security or Sustainable Food Supply System

Following are the key features to ensure availability of healthy food to meet the diet requirements of all the individuals:

(i) Sustainable Agriculture

Integration of traditional and modern agricultural techniques ensures following benefits:

- ☆ It helps minimize negative environmental impacts.
- ☆ It minimizes over-exploitation of natural resources *i.e.* soil and water.
- ☆ It maintains the quality and safety of food.
- ☆ It provides employment opportunities to rural farmers.
- ☆ It helps self sufficiency of villages by rural development. The concept of developing a self-sufficient village with the help of techniques of vermicomposting, biogas production, biofertilizers, biopesticides *etc.* is known as **biovillage** or **ecovillage**.

(ii) Environmental Awareness

Creating awareness among general public, about nutrition values of food and balanced diet, helps building food security or sustainable food supply system.

2.1.2 Challenges to Food Security

Following are the obstacles in achieving equitable distribution of food resources:

(i) Population Increase

Population is increasing exponentially, to put pressure on food resources to fulfill their food requirements.

(ii) World Food Problems

World Food and Agriculture Organization (FAO) estimates, a healthy man requires minimum of 3000 Kcal/day and a healthy women requires 2200 kcal/day. Due to unemployment, poverty, bent toward junk food, less availability of fresh food and consumerism lifestyle following world food problems are discussed:

1. Under-nourishment

Lack of availability of food (less than 90 per cent of the minimum calorie requirement) due to poverty *e.g* Kwashiorkar and Marasmus in children due to lack of proteins and calories.

2. Mal-nourishment

Lack of balanced diet (nutrients), in the intake of food due to urban lifestyle *e.g.* liking for the junk-food (**Table 2.1**) *etc.*

Table 2.1: Some examples of malnutrition causing deficiency of nutrients

Deficiency of Nutrients	*Effects of Health*
Vitamin A	Night Blindness
Vitamin C	Scurvy or Gum disease
Vitamin D	Weakening of bones or Osteoporosis, Rickets
Iron	Anaemia (less haemoglobin: weakness, fatigue, sleepiness)
Iodine	Goitre (Hypothyroidism *i.e.* less active thyroid gland)

(iii) Over-nourishment

Due to over- eating of saturated fats, sugar and salt in the diet, obesity diseases, high blood pressure, diabetes and heart diseases *etc.* are caused, is known as over-nourishment.

2.1.3 Shelter Security

Shelter security means safety, peace and access to food, water, energy, shelter, health, and livelihood. This is an important issue of safety at individual and national and international levels. United nations report has mentioned the issue of shelter security as an important agenda towards sustainable development.

2.1.4 Economic Security

Sustainable development does not merely indicate growth, but a growth in such a way that to make it less material- and energy-intensive and more equitable in its impact *i.e.* less natural resource depletion and a wider perspectives in terms of availability to nearly all the individuals or equitable sharing of natural resources.

2.1.5 Social Security

The social aspects *e.g.* culture, religion, festivals, get-together, function celebrations and variety of amusements are important parts of social systems, which help generate interactions and develop a strengthened self-sufficient system.

2.2 Effect of Human Activities on Environment: Agriculture, Housing, Industry, Mining and Transportation Activities

Table 2.2: Effect of Human Activities on Environment

Mining Mineral extraction from its ore)	**Industries**	**Transportation**	**Housing**
Air pollution Blasting and extraction. Toxic gases and dusts	**Air pollution** Discharge of toxic gases	**Air pollution** Release of gases from vehicles	**Air pollution** (cement dust)
Water pollution Run-off, dust particles and chemicals are reached into water bodies	**Water pollution** Water is used as coolant for machines.	**Marine pollution** Oil leakage	**Pressure on water resources** Water is used in making cement pozollanic.
Noise pollution Heavy machines and transportation	**Noise pollution** Eequipments and machines	**Noise pollution** Horns	**Noise pollution** Drilling and machine processes produce noise
Deforestation For extracting minerals present underground	**Deforestation** For using forest land for setting industries	**Deforestation** For construction of road and railways	**Deforestation** For using forest land for residential purposes
Soil erosion Mining operations remove top fertile layer of soil	**Soil pollution** Solid wastes are disposed in soil.	More cemented area needed for transportation.	Construction activities cause soil erosion.

Contd...

Table 2.2–*Contd...*

Social problems (to local people sanitation, sewage disposal, water supply, routine problem)	Pressure on land and other resources for raw material Pressure on transport system. Problems to local people.	Human health problems.	Extraction of construction materials and energy utilization.

2.2.1 Effect of Modern Agriculture on Environment (*MTU, 2011*)

Ill-Effects of Modern Agriculture (*MTU, 2012*)

In modern agriculture, High Yielding Varieties (HYV) are grown which require large amount of water supply, chemical fertilizers and pesticides. (Green revolution) It causes following environmental problems:

(i) Soil Erosion

Excessive water-supply and wind removes the top fertile layer of the farm known as soil erosion.

(ii) Ground Water Contamination

Leaching of nitrates (NPK fertilizers) may cause ground-water pollution. *e.g.* haemoglobin combines with nitrates to form methaemoglobin. The disease caused is methaemoglobinemia or *blue baby syndrome or cyanosis*. (New born baby are blue in color and stops oxygen supply to body parts).

$$\text{Nitrates in chemical fertilizers} \xrightarrow{\text{leaching}} \text{Ground water (which is drinking water)}$$

$$\text{Nitrates + Haemoglobin} \longrightarrow \text{methaemoglobin} \longrightarrow \text{stops oxygen supply to different body-parts}$$

(iii) Water-Logging and Salinity (UPTU, 2010)

Due to improper drainage, water gets accumulated in the farm known as water-logging. When water evaporates, it causes salinity in the soil. The plants die (**Figure 2.1**).

(iv) Eutrophication (MTU, 2013-14)

The water runoff with chemical fertilizers will reach to the nearby water – body. Small water plants grow in excess known as Algal Bloom. It prevents or stops inter-mixing of atmospheric oxygen to Dissolved Oxygen in water. The water plants and animals start dying due to lack of oxygen. The dead parts will deposit at the bottom of the water body. The process continues and cause threat to the water-body. The process is known as Eutrophication. The water body is known as Eutrophic water body (**Figure 2.2**). *e.g.* Chilka Lake in Orissa.

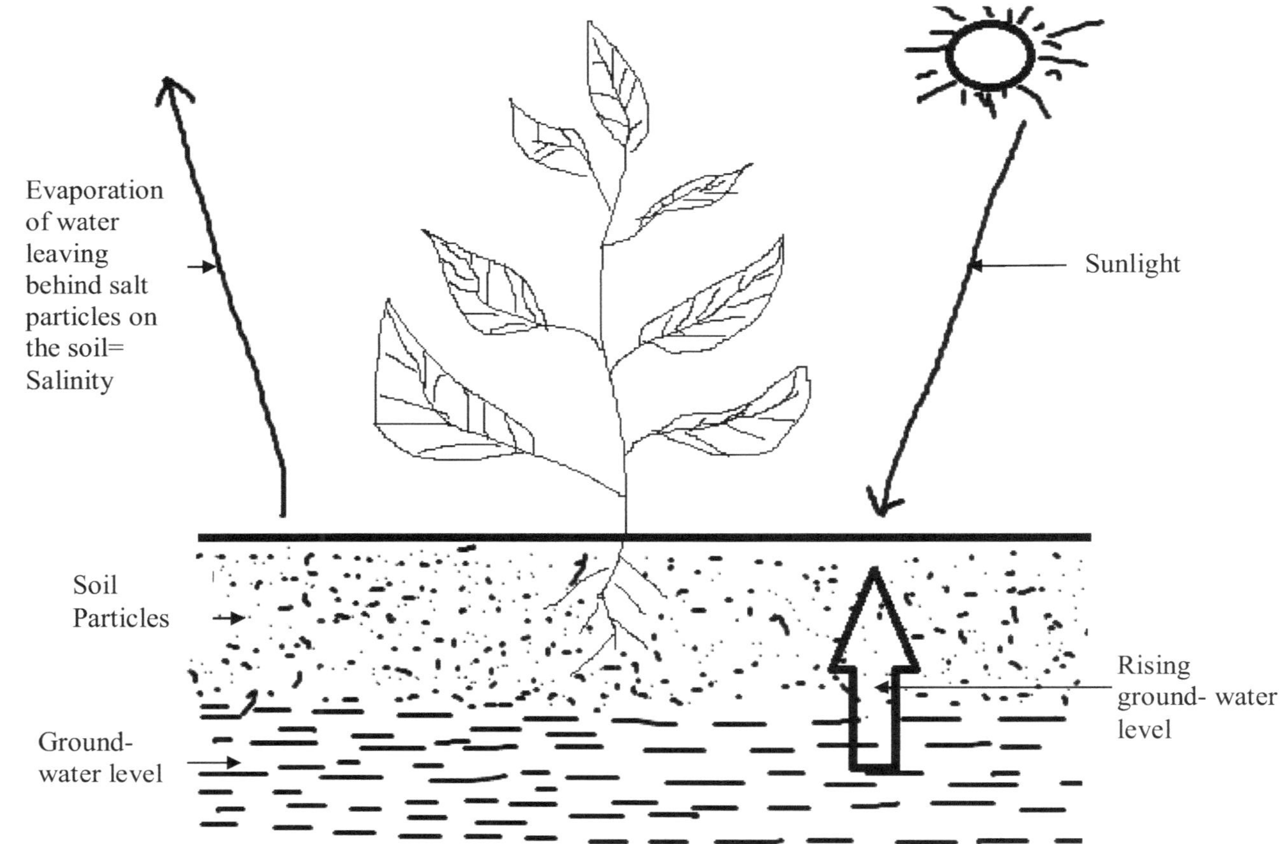

Figure 2.1: Due to Improper Drainage in Indian Farms, Groundwater Level Rises Causing Water-logging and Salinity.

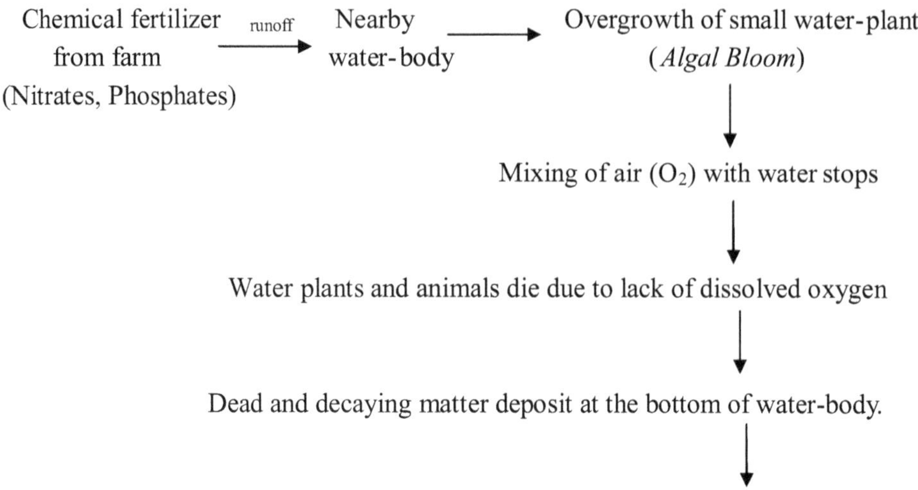

Figure 2.2: The Process of Eutrophication.

(v) Biomagnification

Concentration of the toxic substance increases several times when it is transferred from one organism to another, in the food-chain known as Biomagnification *e.g.* Increase in DDT concentration caused reproductive failure in birds *i.e.* their eggs hatch before time (**Figure 2.3**).

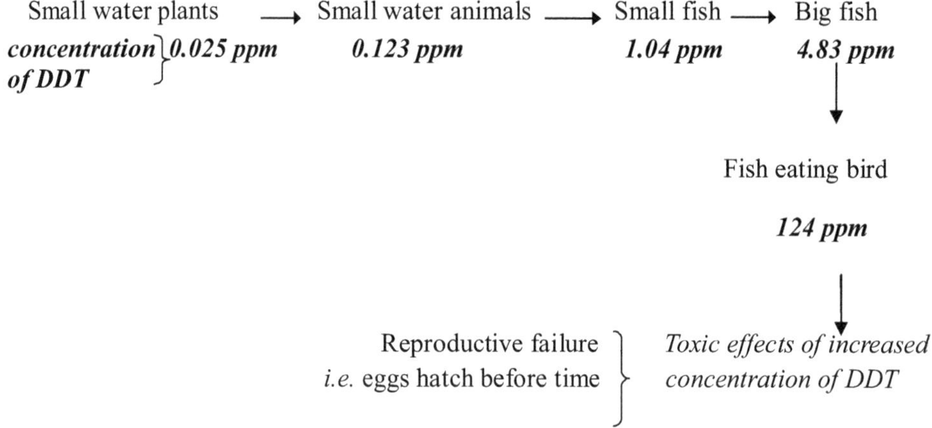

Figure 2.3: The Process of Biomagnification

(vi) Seed Suicide

High Yielding Varieties (HYV) is used in modern agriculture, to produce more crop yield. They require large amount of water and fertilizers *i.e. HYV are irrigation and chemical fertilizer intensive.*

(i) Excessive use of chemical pesticides kills the pests. However growing the same kind of seeds again and again *increases the chances of pest-attack*. Thus the crop becomes vulnerable to pest attack. *e.g.* Potato famine occurred in Ireland in 1969 due to pest attack on the complete crop due to **monoculture**. In other words, we can conclude that seeds of the crops attacked by pests have committed suicide and the farmers also commit suicide due to crop failure.

(ii) Seeds of HYV do not germinate *i.e.* loose their viability every season due to terminator technology being used in their genes. Thus, farmers used to buy these seeds every year.

(iii) Seeds of High Yielding Varieties are very costly. Farmers are compelled to use these high yielding varieties seeds to increase their crop production and to gain profit. Clever advertising by corporate of seeds and chemical industries and *easily available credits (loans)* for the purchase of costly seeds and chemical fertilizers (**Figure 2.4**).

Farmer's Suicide Scenario

- ☆ **Punjab:** Punjab has produced highest rate of farmer's suicide among all the states. The factors causing farmer's to suicide are: crop-failure, monoculture, easy pest attack, seed-failure, decline in farmer's income rate, crop-loss, seed failure *etc.*
- ☆ **Andhra Pradesh:** The farmer's suicide rate increased 20,000 due to worst cotton cultivation, crop failure, and severe pest-attack.
- ☆ **Karnataka:** Chili crop suffered complete destruction in the Bellary region, due to excessive rainfall and viral attacks in 2000. Cold storage facility was also made very costly. Due to the high input cost and deposited loan issues, farmers committed suicide.

(vii) Farmland Conversion

More and more cemented areas are developed as a result of industrialization. Converting a farmland into cemented area is known as farmland conversion.

2.2.2 Effect of Housing on Environment

Increasing population puts tremendous pressure on natural resources to fulfill the growing demands of providing shelter or housing. The process of providing housing to the growing population puts following impacts on environment:

1. **Natural resource depletion:** More construction material is extracted from natural resources *e.g.* minerals, bamboos, bricks, cement, flyash.
2. **Deforestation:** Forest land is cleared for housing purpose and obtaining timber.
3. **Energy utilization:** Energy consumption increases for the purpose of construction and inside the houses.
4. **Water drainage:** Natural water drainage system is disrupted.

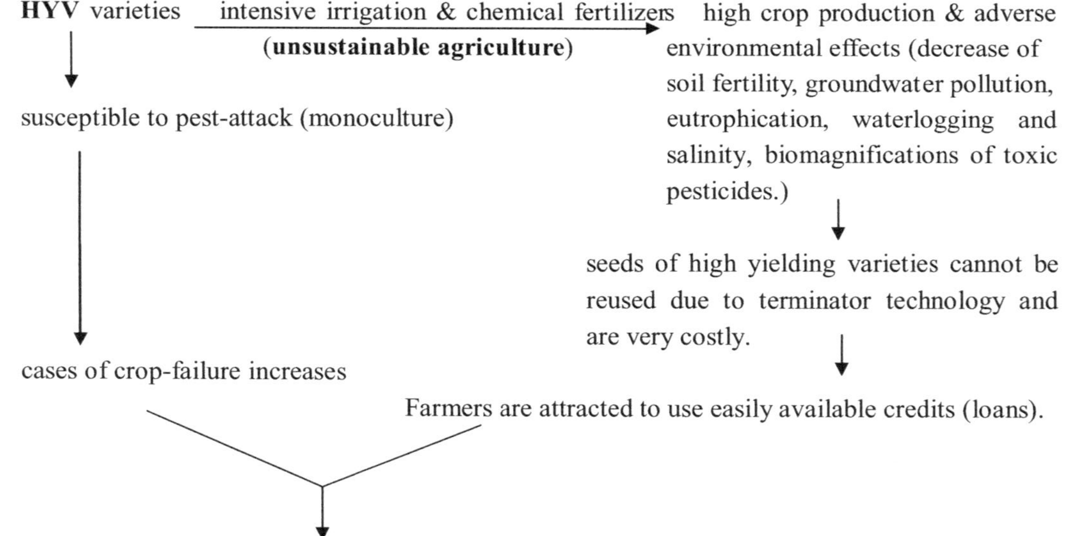

Farmers are compelled to suicide in the worst conditions of crop-failure and loans.

Figure 2.4: Seed-suicide Scenario of Modern Agriculture or Ill Effects of HYVs.

5. **Disposal of discarded waste:** Rejected building materials are to be disposed off either on the wasteland or treated.

2.2.3 Effect of Power Generation on Environment

(i) **Deforestation:** Forest land is cleared for building thermal power plants.

(ii) Burning of coal in thermal power plants cause air pollution.

(iii) **Radioactive pollution** is most probable in nuclear power plants *e.g.* Chernobyl disaster, Fukushima disaster.

2.2.3.1 Effect of River Valley Projects (Water Resource Projects) on Environment

(i) **Deforestation:** Forest land is cleared for dams. Forest land is cut for non-forest purposes. Biodiversity loss occurs at a greater extent.

(ii) **Displacement of people:** Large number of people is displaced due to construction of dams.

(iii) **Water-borne and water-induced diseases:** Breeding of mosquitoes spread vector-borne diseases *e.g.* malaria, dengue.

Reservoir Induced Seismicity (RIS)

Due to the pressure of the water column on the earth core, it increases the probability of earthquakes.

2.2.4 Effect of Mining on Environment (*MTU, 2008*)

Environmental problems of mineral extraction (mining) or ill effects of mining (Table 2.3)

(i) Air Pollution

Toxic gases are release in the atmosphere in the process of mineral extraction.

(ii) Deforestation

Trees are cut to clear the land for mining known as Deforestation.

Table 2.3: Ill effects of Mineral Extraction.

Mineral Extraction or Mining	*Toxic Gases Released in the Atmosphere*	*Environmental Effects*
Cupper (Cu) mining *e.g.* in Khetri cupper mines, rajasthan.	SO_2	***On plants*** (i) Degradation of chlorophyll to brown pigment. (ii) Green plants stops Photosynthesis. ***On animals*** (i) Respiratory problems
Iron (Fe) mining *e.g.* Kudramukh Iron ore mine, Karnataka	CO, CO_2	***On plants*** Green plants stops photosynthesis. ***On animals*** Reduces the ability of haemoglobin to carry oxygen, forming ***Carboxyhaemoglobin.***

(iii) Soil Erosion

Large area of land is spoiled, taking out minerals from soil known as **soil erosion.**

(iv) Water Pollution

The water run-off with unused parts of minerals called mine tailings cause water pollution *e.g.* Gold-mining is done by hydraulic washing and treating it with cyanide solution. Cyanide percolates into the groundwater known as **heap leaching**. It is lethal (highly toxic, cause death) to plants and animals of surface water and ground water.

(v) Noise Pollution

Heavy machines are used in mining produce loud noise.

(vi) Landslides

Movement of lands may occur.

(vii) Unemployment after Finishing of the Mineral Reservoir

Case Study of Khetrinagar Copper Project

Copper extraction from its ore present in the rocks has been carried out in the Khetrinagar cupper project. All the extraction of cupper has been done *i.e.* the rocks are now finished with the cupper ores. This indicates the non-renewable nature of the minerals. The whole project is now under final completion, leading to unemployment.

2.2.5 Effect of Transportation Activities on Environment

Transportation industry has following impacts on environment:

(i) **Deforestation:** Forest land is cleared for building roadways, railway tracks, and airports.

(ii) **Disposal of discarded waste:** Rejected materials used in construction and wear and tear of vehicles are to be disposed off or recycled.

(iii) Automobiles are the largest source of air pollution and also cause noise pollution.

(iv) **Natural resource consumption:** Production of automobiles puts pressure on mineral resources *viz.* metals

2.2.6 Effect of Tourism on Environment

Hotspots of biodiversity (areas with very rich biodiversity) are the main centre for attraction of tourism. They are also the conservation priority areas. Thus biodiversity faces danger of extinction if the tourism increases in their habitats due to disturbance in their physiology, overcrowding, dispersal of plastic wrappings and other solid wastes, noise, and pollution due to transportation.

The tourism integrating with environmental conservation is known as ecotourism *i.e.* principles of environmental conservation along with benefits of tourism.

Participatory Learning

(a) Objective Questions

1. Overgrazing results in ______. (*UPTU, 2006*)
2. HYV require large amount of ______ and ______.
3. India is one of the ______nations.
4. Cholera is a ______disease.
5. World food day is celebrated on ______.
6. Extensive irrigation under improper drainage leads to______.
7. ______ is the primary adverse environmental effect.

(b) Very Short Answer Type Questions

1. What is seed-suicide?
2. What is reservoir induced seismicity (RIS)?

(c) Short Answer Type Questions

1. Write short-notes on (***UPTU, 2010***)

   (a) Water-logging (b) Salinity (c) Eutrophication.
2. Describe the ill effects of automobile pollution.
3. Define soil erosion. How it is caused?
4. What do you understand by the concept of biovillage or ecovillage?

(d) Long Answer Type Questions

1. What are the ill effects of modern agriculture? (***MTU, 2012***)
2. Explain the ill-effects of mineral extraction. (***UPTU, 2010; MTU, 2013-14***)
3. What are the adverse impacts of automobile pollution?
4. What is meant by green revolution and what has been its impact? (***UPTU, 2010***)

Answers to Objective Question

1. Soil erosion
2. Water and chemical fertilizers
3. Mega-diversity nations
4. Water-borne disease
5. 16, October
6. Waterlogging and salinity
7. Desertification

Chapter 3

Natural Resources

Natural Resources (*UPTU, June 2009*)

The materials useful to man in nature's supply or reservoir are known as natural resources. e.g. air, water, land, forest, crops.

3.1 Types of Natural Resources

Renewable Natural Resources

Materials present in nature's reservoir which are renewed or refreshed within a short time-period. *e.g.* geothermal energy, tidal energy and hydrogen.

Non-Renewable Natural Resources

Materials present in nature's reservoir which take a very long time in their formation. *e.g.* minerals. Their rate of consumption is very high as compared to the rate of formation.

Perpetual Natural Resources

Materials present in nature's reservoir, which are inexhaustible or their supply is continuous. *e.g.* solar energy, wind energy and water.

Sun as a Natural Resource

Green leaves receive sunlight and make their food by the process of photosynthesis in the presence of carbon-di-oxide (CO_2), water (H_2O) and sunlight.

3.2 Natural Resources Availability and Problems or Natural Resource Crisis and Population Explosion

Population is increasing exponentially (*i.e.* at a very faster rate), puts pressure on Natural Resources (*e.g.* air, water, soil, and other energy resources), to fulfill the growing demands of population *i.e.* food, clothing and shelter, known as **natural resource crisis** or **natural resource crunch**. *e.g.* depletion of ground-water level, depletion of coal and petroleum (their rising cost). The developmental activities also consume our important natural resources. Population increase and natural resource crisis together is known as **population explosion (Figure 3.1**).

Population increase (exponential in nature)

⇩

puts pressure on natural resources (*e.g.* water resources)

⇩

natural resource crisis or natural resource crunch or depletion of natural resources
(*e.g. ground-water depletion*)

Figure 3.1: Crisis of Water Resources or Adverse Effects of Withdrawing Excess of Groundwater (*MTU Carryover, 2012-13*)

3.3 Water as a Vital Natural Resource

- ☆ Water is required by green plants. They carry out photosynthesis in the presence of CO_2, water and sunlight. Crops are irrigated with water.
- ☆ Hydroelectricity is produced from water.
- ☆ All industrial activities require water as a coolant and cleanser.

3.3.1 Water Cycle (Hydrological Cycle)

The cycling of evaporation of water and consequent rainfall helps maintain water-cycle (**Figure 3.2**).

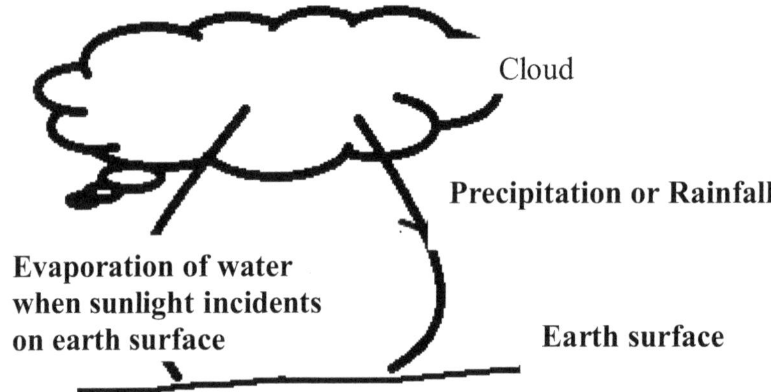

Figure 3.2: Water-Cycle or Hydrological Cycle.

3.3.2 Availability and Quality Aspects (groundwater depletion) *(MTU, Carry over, 2012-13)*

Due to increasing population, the demands of water supply increases, putting pressure on water resources, leading to ground water depletion (**Figure 3.3**).

Population increase (exponential in nature)

puts pressure on water resources

water resource crisis (*e.g. ground-water depletion*)

Figure 3.3: Water Crisis.

3.3.3 Water-borne and Water-induced Diseases

The diseases caused by coming in direct contact with the contaminated (dirty) water *i.e.* drinking contaminated water. Their causative agents and symptoms are given below (**Table 3.4**).

Table 3.4: Water-borne Diseases.

Water-borne Disease	*Disease Causing Agent i.e. Pathogen*	*General Symptoms*
Typhoid	Bacteria: *Salmonella typhii*	Long-term fever, sweating, diarrhoea, liver enlargement, finally death.
Amoebiasis (Hand-to-mouth)	Protozoa: Amoeba	Stomach pain, fatigue, weight loss, diarrhoea, fever.
Taeniasis	Tapeworms	Intestinal disturbances, loss of weight.
Cholera	Bacteria: *Vibrio cholera*	Very dangerous disease, very watery diarrhoea, nausea, vomiting. It may lead to death.
E. Coli infection	*Escherichia coli* (*E. coli*)	Mostly diarrhoea can cause death due to dehydration.

Water-induced Diseases

Diseases caused by indirect contact with the contaminated water are known as water-induced diseases (**Table 3.5**).

Infection from dirty water $\xrightarrow{\text{carrier}}$ reaches to the blood of the patient

Stagnant water → mosquito → bite to a human → water induced disease

Table 3.5: Water-induced Disease.

Water-induced Disease	*Disease Causing Agent i.e. Pathogen*	*Carrier-agent/ Vector*	*General Symptoms*
Malaria	*Plasmodium* sp.	Female anopheles mosquito	Fever with shivering, sweating
Dengue	Virus	Mosquito	Fever

3.3.4 Fluoride Problem in Drinking Water (*UPTU, June 2010*)

Due to intake of excess of fluorine in drinking water, it interferes with Calcium metabolism, known as fluorosis. This results in weakening of bones and teeth.

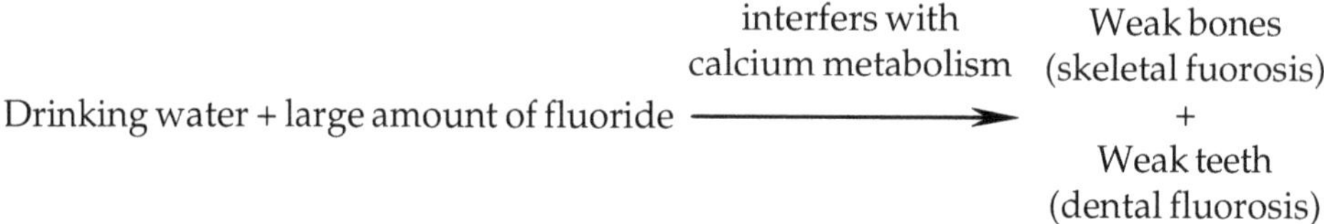

Salient Features

It is an endemic disease *i.e.* restricted to a particular area. *e.g.* in Rajasthan in India, concentration of fluoride in ground water is naturally very high.

Types of Fluorosis or Symptoms of Fluorosis

Skeletal Fluorosis

Weakening of bones.

Dental Fluorosis

Weakening of teeth.

Measurement of Fluoride

Fluoride content in drinking water is measured by Ion Selective Electrode (ISE) Meter or Spectrophotometer.

Importance of Fluoride

Fluorides are naturally found in water at low concentrations (1mg/l); it is beneficial/good for teeth. Thus low concentrations of fluoride are added in the tooth pastes known as fluoridation

Permissible limit of fluorine in drinking water is 1.5 mg/l as per IS: 10500 standards. Regular consumption of water with more than 8mg fluoride/liter of water causes fluorosis.

Process to Remove Fluorine from Drinking Water (Defluoridation)

(i) Nalgonda Technique (UPTU, Dec 2009)

It is an Indian traditional technique, to remove fluoride ions, from water with the help of alum. Water sample is treated or shaked with alum and lime, a precipitate of fluoride is removed. It is a defluoridation technique given by a person "Nalgonda" in an Indian village.

$$\begin{array}{c}\text{Water sample}\\ \text{(containing fluoride beyond}\\ \text{permissible limit)}\end{array} + \text{Alum} + \text{Lime} \xrightarrow{\text{shake}} \begin{array}{c}\text{Fluoride precipitates} \downarrow\\ \text{(removal from water)}\end{array}$$

The technique is recommended by NEERI (National Environmental Engineering Research Institute, Nagpur) for removal of fluoride.

(ii) Activated Carbon treatment.

(iii) Fluoride exchangers using bone charcoal.

(iv) Treatment with lime and magnesium salts.

(v) Treatment with aluminium salts and coagulants.

(vi) Reverse Osmosis.

3.3.5 Arsenic Problem in Drinking Water

Arsenic contaminated drinking water causes **Arsenecosis** or **Black foot disease**. Arsenic contamination of drinking water occurs due to excessive use of arsenic containing chemical pesticides *viz.* cupper arsenate and lead arsenate in the modern agriculture.

$$\text{High levels of arsenic} + \begin{array}{c}\text{Groundwater}\\ \text{(drinking water)}\end{array} \longrightarrow \textbf{Arsenecosis}\text{ or }\textbf{Black foot disease}$$

WHO Recommendations

(i) Safe Levels of Arsenic in Drinking Water = 10 μg/litre, Now recently recommended Arsenic Safe Level is 0.01 μg/litre

(ii) Upper Limit of Arsenic in Drinking Water = 50 μg/litre.

Symptoms

The patients of arsenecosis suffer from following problems:

(i) Black spots on chest, back, limbs (hands and legs) known as melanosis.

(ii) The skin becomes hard and fibrous.

(iii) Severe toxicity can lead to Gangrene and Cancer.

(iv) Complications of liver, spleen, goiter and skin cancer may also develop due to arsenic poisioning.

Areas Reported with Arsenic Toxicity

India: Several cases of Arsenecosis have been reported from West Bengal and Bihar (India).

World: Bangladesh has some of the most polluted ground water in the world. 85 per cent of the area of Bangladesh's ground water is arsenic contaminated. 1.2 million people of the country of Bangladesh is suffering from arsenic poisioning.

Control of Arsenecosis

(*i*) Government analyzes water samples for arsenic contamination and mark the arsenic contaminated tube-wells with red paint and tube-wells containing safer water are marked with green paints.

Bioremediation

(*ii*) A fern plant *Pterris* sp. removes arsenic from polluted water.

3.4 Minerals as a Natural Resource

Minerals are used in agriculture, defense, and transport, industrial and other sectors (**Table 3.6**).

Table 3.6: Most Common Minerals and their Uses.

Minerals	*Uses*
Aluminium (Al)	Air-crafts, electric cables, utensils.
Copper (Cu)	Electric industry as wires.
Gold	Ornaments

3.5 Forest as a Natural Resource

- ✰ Forest is the group of variety of trees, which support growth of variety of plant and animal species.
- ✰ Variety of ecological (air purification, water-cycle, pollination, prevents soil erosion, habitat of many animals) and economic uses (fodder, fuel, fibre, fruit, sandal wood, medicines, resins *etc.*) of forests exists (Figure 12).

  Forest=Trees + Herbs + Shrubs + Grasses + Animals + Insects + Microorganisms
- ✰ National Forest Policy recommended 33 per cent forest area for plains and 67 per cent forest area for hills.

3.5.1 Over-Exploitation of Forest Resources

Forest provides variety of timber (*e.g.* wood) and non-timber based products (*e.g.* fibre, fodder, medicines, resins, oils) to the increasing population (**Figure 3.4**).

Population increase (exponential in nature)

Puts pressure on forest resources

Depletion of forest resources to fulfill the demands of increasing population (*viz.* fruit, fibre, fodder, furniture, timber, medicines, resins, oils).

Figure 3.4: Over-Exploitation of Forest Resources.

3.5.2 Conservation of Forest Resources

Due to excessive exploitation of forest resources, there is a need to protect forest area from deforestation. Conserving forests helps in sustainable development. In-situ and *ex-situ* conservation measures of biodiversity are described in Chapter "Biodiversity". Environmental legislation to protect forest area is Forest conservation act, 1980 and Wildlife Protection Act, 1972.

3.6 Food Resources

3.6.1 Conservation of Food Resources: Sustainable Agriculture (Organic-Farming) or Solution to Modern Agriculture

Integration of traditional agricultural techniques into the modern agriculture is the concept of sustainable agriculture. List of important techniques of sustainable agriculture are given as follows:

(i) Agroforestry

Planting trees in agricultural farm has following benefits (**Figure 3.5**) *e.g.*

- ☆ Roots of the trees keep the soil intact *i.e.* reduces soil erosion.
- ☆ Products from the trees provide economic benefits to farmers in the season of no harvesting.

(ii) Biopesticides

Preferable use of biological origin based pesticides in place of chemical pesticides *e.g.* Neems.

(iii) Biofertilizers (or green manures)

Naturally occurring living organisms used as fertilizer is known as biofertilizer. They supply multiple nutrients to plants by various metabolic transformations; in easily absorbable form. Faba bean, Broad bean (*Vicia faba*), Common bean, Kidney bean (*Phaseolus vulgaris*), Cow pea (*Vigna unguiculata*), Pigeon pea (*Cajanus cajan*), Jack bean (*Canavalia ensiformis*), Lima bean (*Phaseolus lunatus*), Mung bean (*Vigna radiata*), Soyabean (*Glycine max*), Rice bean (*Vigna radiata*), Velvet bean (*Mucuna pruriens*), Winged bean, Asparagus pea (*Psophocarpus tetragonolobus*), Sunhemp (*Crotolaria juncea*) *etc.* are important green manures.

Naturally occurring living organisms *e.g. Vicia faba, Phaseolus vulgaris* $\xrightarrow[\text{metabolic transformations}]{\text{supply and absorption}}$ Group of essential nutrients to plants

(iv) Scare-crow

It is a human-dummy, placed in the farm to protect the crop from birds.

(v) Mixed-Cropping

Group of many varieties of a crop were grown in a farm. It helps reduce the possibility of pest-attack by confusing the pest. Growing same type of crops (High Yeilding Varieties) in the farm increases the chances of pest-attack because the pest develops resistance to the same crop is known as monoculture. However, traditionally growing variety of crops in the same cropland reduces the chances to pest attack

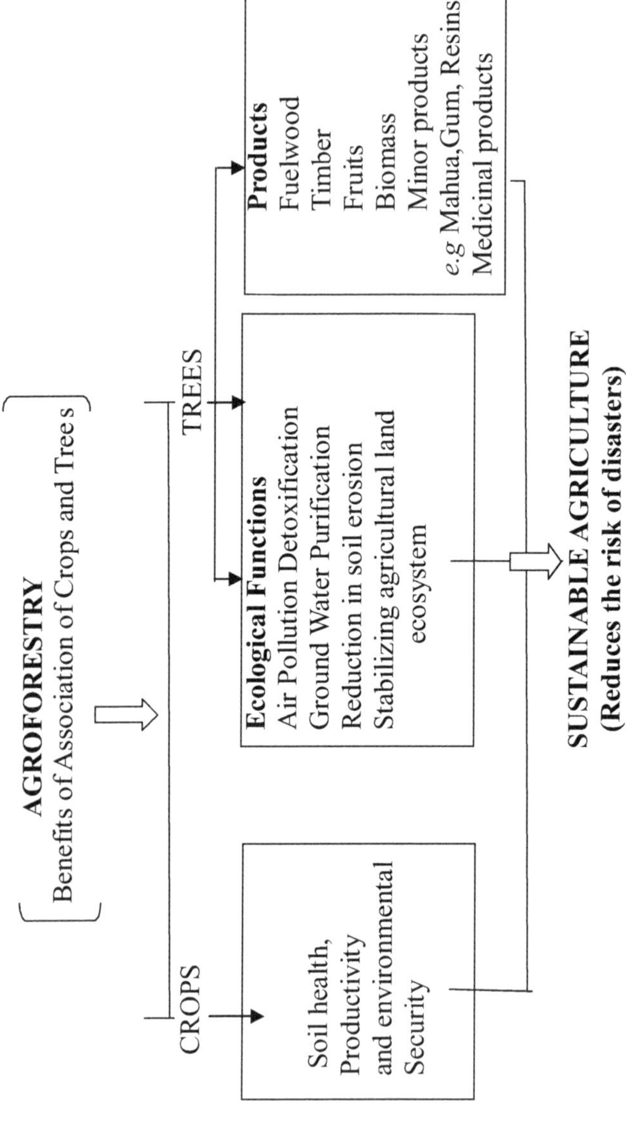

Figure 3.5: Agroforestry Approach.
(*Source*: Ecointrospection of sustainable agricultural techniques in India, D. Agarwal *et. al.*, in Sustainable Development, Anmol publications)

because the pest machinery gets confused (in variety of crops). *e.g.* Potato famine occurred in Ireland (in 1964).

(vi) Crop-rotation

Group of crops are grown in rotation in the farm so that the nutrients consumed by one crop are been supplied by the other crop. *e.g.* Wheat consumes Nitrogen from the soil and pulses replenish Nitrogen (Rhizobium bacteria in the root-nodules of leguminous plants fixes atmospheric nitrogen to usable form by the process of **biological nitrogen-fixation**), thus replenishes nitrogen in the soil (**Figure 3.6**).

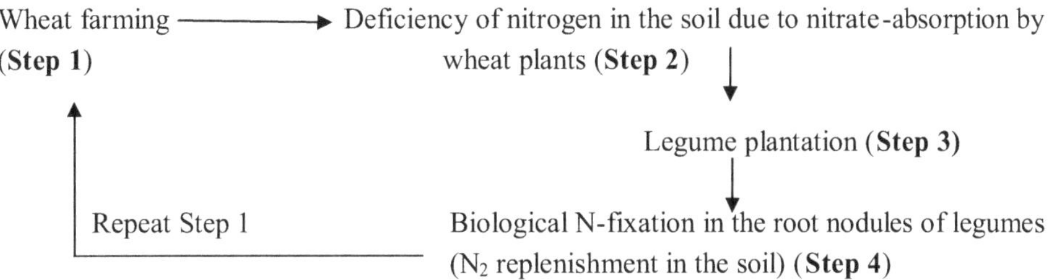

Figure 3.6: Process of Crop-rotation.

(vii) Innovative Techniques

Mustard is grown in the wheat farm. The smell of mustard repels insects. It protects the crops from pest-attack.

(viii) Terrace Farming or Step Farming

In hills crops are grown in steps to stop soil erosion.

(ix) Ploughing

Domestic animals used for ploughing in the farm also play a key role in self-sustenance of closed nutrient cycling system in the farm. Animal wastes enrich the soil with micro and macronutrients. Wastes from cropping systems are used as food for animals.

(x) Mulching (*MTU, 2013-14*)

Inorganic (plastics, rock chips *etc.*) and organic matter (straw, compost sawdust) is used to cover the ground after the crop is harvested. It protects the plant roots from heat, cold or drought stress and also reduces soil erosion.

3.7 Land as an Important Natural Resource

Soil is a rich source of minerals for human usage and nutrients for plant growth. Maintaining its productivity is an important feature of sustainability.

3.7.1 Land Degradation

Decrease of fertility of soil due to man-made activities is known as land degradation.

3.7.2 Wasteland Reclamation (Restoration)

Planting trees on wasteland is the most viable environmental option for regaining the productivity of wastelands. *E.g.* planting tolerant trees *i.e. Arjun, Neem, Peepal, Bargad* are tolerant to heavy metals.

Participatory Learning

(a) Objective Questions

1. Which of the following is natural resource?

   (a) Air (b) Soil

   (c) Water (d) All of these

2. Which of the following is not natural resource?

   (a) Forest (b) Minerals

   (c) Biodiversity (d) CFCs

3. Major purpose of dam is

   (a) Flood control (b) Hydroelectricity

   (c) Fishing (d) Irrigation

4. Excessive intake of fluoride, causes ______. (***UPTU, 2010***)

5. Most biotic resources are ______. (***UPTU, 2009***).

6. World water day is observed on______.

7. Khetri copper project is famous for______ in______.

True/False

8. Cauvery water dispute is between Karnataka and Tamil Nadu.

(b) Short Answer Type Questions

1. What are perpetual, renewable and non-renewable natural resources?

2. What are conventional and non-conventional natural resources?

3. Does bioremediation play important role in decontamination of water?

4. What percentage of water is present in oceans and earth's surface?

5. Define minerals. What is mine spoil?

6. What are occupational diseases?

7. What are the ill effects of deforestation?

8. Sunderlal Bahuguna is famous for which environmental reason?

9. Give the measures which serve as solutions to deforestation?
10. What is Chipko movement, Appiko movement and Bishnoi movement?
11. Describe the taungya system of agroforestry?
12. What do you understand by Jhoom cultivation or splash and burn agriculture?
13. How much forest area is covered by the National Forest Policy (1988) for plantation in hills and plains?
14. Give the importance of women in environmental conservation. Who is the first woman involved in Bishnoi movement?
15. What is land degradation?
16. What is desertification?
17. Define pedology and pedogenesis.
18. What is soil erosion? Give its types.
19. What is weathering?
20. What is organic farming?
21. What is sustainable agriculture? Does it contributes to combat climate change?
22. What is heap leaching?
23. What are water-borne and water-induced diseases? **(*MTU, 2012*)**
24. What is eutrophication? **(*MTU, 2013-14*)**
25. What is mulching? **(*MTU, 2013-14*)**

(c) Short Answer Type Questions

1. What are the methods of conservation of water? Define water-shed management and rain-water harvesting. **(*UPTU, 2006, 07; MTU, 2012-13*)**
2. Define aquifer. Give its types.
3. Describe the fluoride problem in drinking water. Give its types. Explain nalgonda technique. (***UPTU, 2009***).
4. Explain the defluoridation techniques of fluoride water.
5. Describe causes, effects and control measures of arsenic problem in drinking water.

6. Write short note on rain water harvesting and water-shed management.
7. What are the measures to combat droughts and floods?
8. How arsenic can be removed from water?
9. Describe the distribution of water in various environmental segments.
10. Explain the benefits and problems of river valley projects or dams.
11. What are the ill effects of mineral extraction or mining? (***UPTU, 2008, 2009***)
12. What is the need for conservation of minerals? (***UPTU, 2009***)
13. Comment "water as a natural resource".
14. Comment "mineral as a natural resource".
15. Comment "forest as a natural resource".
16. Comment "biodiversity as a natural resource".
17. Comment "land as a natural resource".
18. Comment "soil formation takes a very long passage of time".

(d) Long Answer Type Questions

1. Describe the crisis and conflicts over water resources (***UPTU, 2009; MTU, 2012***)
2. Describe the various water-quality standards or parameters of water quality.
3. How the excess of water-supply in the farm adversely affects crop-production (***UPTU, 2011***)
4. Water body ids polluted, lot of waste is scattered here and there. Explain how it is caused? What are the effects and control measures? (***UPTU, 2008, 2009***)
5. Explain the waste water treatment plant lay-out. Explain the steps. (***MTU, 2013-14***)
6. Define thermal pollution. What are the causes effects and control measures of thermal pollution? (***UPTU, 2011***)
7. Explain the steps of soil conservation.

Answers to Objective Questions

1. (d)	2. (d)	3. (d)
4. Fluorosis	5. Renewable	6. 22 March
7. Copper mines, Rajasthan	8. True	

Chapter 4

Biodiversity

4.1 Definition of Biodiversity

Variety or variability among living organisms, and variety among the ecosystems in which they live, is known as biodiversity.

Bio=Life/Living Beings/living Factors, Diversity=Variations/Variety

Article-2 of Convention on Biodiversity defines: "The variability among living organisms from all sources (ecosystems and ecological complexes). This includes variability within species, between species and of ecosystems."

Biodiversity is thus, explained as variety of plants and animals (humans, micro-organisms) and interactions between them *i.e.* sum total of all plants and animals (including humans), fungi and micro-organisms along with their individual variation and interaction between them (**Figure 4.1**).

The term "Biodiversity" is coined by Walter G. Rosen in 1965 in the national forum on Biodiversity held in Washington D.C. in 1986. Biodiversity is now an international buzzword through its use in the Biodiversity Convention, which has been agreed on by the United Nations Conference on Environment and Development, after long debates. (Rio de Janeireo, 1972) and defined as the variability among living organisms from all sources including terrestrial, marine and other aquatic ecosystems and ecological complexes of which they are a part, this includes diversity within species and ecosystems. But as a matter of disappointment to environmentalists, it has being catapulted into the worldwide environmental politics recently for obtaining short-term gains. Biodiversity is the contraction of biological diversity. **(Biodiversity=Biological Diversity)** Term biological diversity is given by Thomas Lovejoy, 1980. Father of Biodiversity is E.O.Wilson.

According to Flint (1991), the 'biodiversity' is the degree of variety in nature, can be understood by a expanding the term as biological diversity. It has come in common

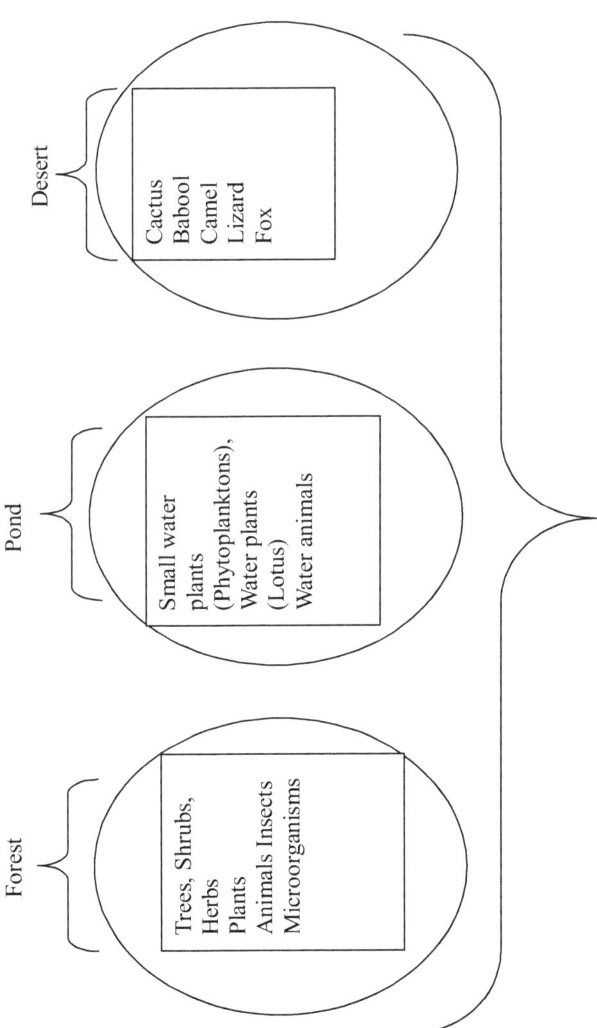

Figure 4.1: Diagram Illustrating the Biodiversity.

usage since around the middle of the 1980's. Biodiversity is the variety and variability among living organisms and the ecological complexes in which they occur. Biodiversity is the degree of nature's variety, including both the number and frequency of ecosystems, species and genes in a given assemblage.

Biodiversity is a very wide term representing genetic diversity, species diversity and ecosystem diversity. It is also described as "The varieties of living organisms at all levels genetic varieties belonging to the same species, family and genera, population, community, habitat of even ecosystems." It refers to the "The wealth of life on the earth, the millions of plants, animals and micro-organisms, the genes they contain and intricate ecosystems they help to build into the living environment."

Thus biodiversity is the contraction of the term biological diversity. To make full utilization of taxa, one should know ecological diversity in terms of species diversity, genetic diversity and species diversity.

4.2 Hierarchical Classification/Origin of Biodiversity

Genetic Biodiversity

The variations in the individuals of single species forms genetic diversity due to wide differences in the genetic constitution. Wide variety of genes are responsible for the degree of variation in color of skin and hairs, size, weight, texture of skin and hairs, height, color of eyes, thickness of nails *etc.* Genes are the basic level of biodiversity wherein the variation in combinations or sequences of the four base-pairs as the components of nucleic acids (DNA), constitute the genetic code and express the form and function in all living organisms.

Nucleic acids (DNA) are contained in each cell in the structures called chromosomes, so alternatively the amount of DNA within the cell may be taken as a measure of genetic diversity. The biodiversity among the same species *i.e.* variation between closely related individuals, with the same chromosome structure and numbers and the same amount of DNA within the cell, needs to be measured by examining the differences in the number and combinations of alleles present. The product of the genetic information contained in the alleles. Simply we can say looking at the chromosome structure or number could identify genetic diversity.

In the traditional farming systems, realizing the importance of variety of genetically diverse plant species, they were promoted and conserved. Total gene-pool is to be maximally utilized *i.e.* interrelate the gene-pool to assess the diversity in totality. For example, more than 5,000 types of old rice variety were cultivated by local tribes: shakkar-cheene, basmati *etc.* Wide gene pool was associated with the variety of cucurbits in ancient Indian subcontinent. Therefore, the taxonomists within the length, breadth of ecological range collectively analyze, assess and conserve biodiversity.

It is the structured variation in genetic diversity responsible for the ability of population to survive. The remaining populations of the cheetah, for example, have been found to be almost identical genetically emphasizing the need of increase in the genetic diversity of the cheetahs rather than just increasing the number of cheetahs that exist. So it is concluded that conservation of genetic diversity carries greater importance.

☆ Variation of genes within the individuals of single species.

Human beings (*Homo sapiens*) is one species.

Difference of genes in different human beings.

One human is different from other in size, shape, colour, height, behaviour *etc*.

☆ Variation of genes occur in following ways:

(i) Variation in the amount of DNA per cell.

(ii) Variation in the structure of chromosomes (DNA).

(iii) Variation in number of genes.

Properties shown by different genes in each species of human beings are size, shape, height, skin, texture (dry skin, oilly skin, normal skin), colour, behavior, boldness, aggressiveness, politeness, resistance against diseases (immunity) *e.g.* diabetes, thyroid, kidney diseases, resistance against adverse conditions *e.g.* hot and cold weather, colour of eye ball.

Species Diversity

Variations among the species in ecosystem i.e. different types of species exist.

A species is biologically defined as a group of populations which can interbreed freely among themselves that is reproductively isolated from other such groups. So the species are sometimes termed as 'biological species'. However, it has been observed that individuals of one species can inter-breed with members of another closely related species. This is actually the magic of nature to maintain the complex balance in nature, *i.e.* biodiversity. The phrase 'change is the law of nature' is true in all senses. *Beauty of nature lies in avoiding monotony*. Natural phenomena keep evolving to produce diversity in the biosphere. So the biodiversity is reflected by variety of morphological, physiological and genetic features.

Each organism carries a specific role of ecological importance. For example, green plants photosynthesize in the presence CO_2, water, and sunlight and prepare their own food. This food energy then gets transferred to herbivores, carnivores, omnivores as the food web continues.

The origination and extinction of species are the key aspects of evolution. Species generally originate when physical or behavioral barriers separate populations of the same species to the extent that they no longer interbreed and at this point they can be described as different species because they are no longer able to maintain a viable population size to the persisting environmental factors and eventually the last representative of the species dies without reproducing. It is thus interpreted that environmental or ecological factors play an important role, in the extinction and formation of species.

Importance of species diversity can be discussed in terms of variety of individuals of different species growing in the biosphere. The phenomenon of endemism due to which the species are restricted to a particular area, adds to the biodiversity. Endemic species are rarely found, so highly important than those which are quite commonly found in other areas as well. Similarly, importance could he given to species, which come from different taxa, organisms of different size classes or organisms representing different trophic (relating to their position in food chains *etc.*) levels. The criteria chosen will depend on the global or local perspective involved.

Certain ecosystems exist only due to the presence of one or a few keystone species (species which perform functions vital to the existence of the ecosystem) *e.g.* Species of fig tree provides fruit to many birds and is a habitat to monkey at the season of leaf fall.

- ✰ Number of different type species in an ecosystem is **species richness**. *e.g.* Many flowering plants, long grasses, rabbits, deers, cows, giraffes, snakes, eagles, birds are different species found in grassland ecosystem.
- ✰ 10-30 million species exist in world. 49,219 (approx 50,000) plant species exist in India.
- ✰ 89,317 (approx 90,000) animal species exist in India.167 crop species and over 350 species of wild relatives of cultivated crops exist in India.

Definition of Species

Individuals which can freely interbreed among each other in an ecosystem are known as species. Species form the basic unit of life.

Ecosystem Diversity

Variety of ecosystems exists, formed by variety of interactions between living and non-living factors. Variety of habitats (natural homes of animals) e.g. Different types of ponds, rivers, lakes, oceans, wetlands, grassland, forest, desert represent different types of living factors and their interactions.

Ecosystem diversity could be best understood if we look at functioning in various ecosystems. Thus, ecosystem diversity includes the variety of species and ecological processes (nutrient-cycling and energy flow) that occurs in different physical settings. In other words, the various components of an ecosystem interact (various adaptations, competition, prey-predator relationship and other interactions like mutualism *etc.*) in a way that they form an integrated relationship and a self-dependent unit. One advantage of considering biodiversity at the ecosystem level is that elements of biodiversity such as the age structure of populations, the pattern of communities within a region, the changes in community composition and structure over time, models of predation, models of parasitism, models of mutualism, *etc.* are not ignored.

Conclusion

Nature has really done a miraculous job by putting variety of plants and animals in different directions. Each of them has been made different. The individuals of a single species are all different among themselves that is actually because of differences in their genetic make-up *i.e. genetic diversity.* We observe none of the rose flowers identical. The shape, size, colour, texture, fragrance, height of the plant, nodules on the stem, internodes, leaf size *etc.* are all different of the rose plants. Major differences have been evolved among different species; they are so different among themselves that they cannot interbreed freely known as reproductive isolation. But they undergo competitions and variety of adaptations to be suited to the climate. The variety of interactions, among these species and climate cause self-sufficient functioning mechanism, in the biosphere.

4.3 Significance of Genetic Biodiversity

Most of the biological diversity arises as a side product of evolution. Natural selection operates in the available gene pool. Thus, the more the genetic variation, higher is the probability for selection of better equipped genotypes, thereby higher is the pace of change in gene frequency leading to wider variations in structure, function and behavior of species in the biosphere.

Species diversity is measured in terms of species number or species richness. However, relative abundance is also important. The number of species in the world is usually estimated to be somewhere between 5 and 30 million, of which only less than 2 million have so far been described.

Biodiversity extinction is a natural process governed in an evolutionary fashion. Natural selection operates in the available gene pool. Thus, the more the genetic variation, higher is the probability for selection of better equipped genotypes, thereby higher is the pace of change in gene frequency leading to wider variations in structure, function and behavior of species in the biosphere which is actually 'Biodiversity' described in the flow chart (**Figure 4.2**).

Variations in genes in the individuals of a species

Ability to adapt to environment, increases

Higher adaptation

Better survival

Figure 4.2: Greater is the Genetic Diversity of a Species, Higher is its Efficiency to Adapt to the Eenvironment.

A series of such complex steps (natural selection, variety of interactions, adaptations, competition *etc.*) lead to higher degree of complexities in the biosphere, forming biodiversity. Here the fundamental evolutionary force is the pace of change in the gene frequency (thereby chromosome configurations) in a population due to the continuous force of natural selection governed by environmental factors. Thus, it can be stated that most of the biological diversity arises as a side product of evolution. Each species is dependent on other within the community. Various mutualistic relationships help their survival. At the same time, competition also exists. The world conservation strategy (IUCN, UNEP, WWF, 1991) statement "conserving biological diversity equals conserving ecosystems".

4.4 Measurements or Measures of Biodiversity

(i) Alpha-diversity (Dinna-diversity or within-community diversity)

Variations of organisms living in the same community or habitat, is known as alpha diversity.

It represents species richness and species evenness.

Species richness= number of variety of species per unit area.

Species evenness= number of individuals of different species.

(ii) Beta-diversity

Rate of migration or displacement of species because of change of environmental conditions, is known as beta-diversity. *i.e.* change in the species composition in an area exists due to environmental gradient.

(iii) Gamma-diversity (=alpha diversity + beta diversity)

Variations of communities or habitats over the large geographical area or landscape, is known as gamma-diversity.

Whittaker (1972) has described three categories of biodiversity: alpha-diversity (diversity within ecosystems), beta-diversity (diversity between ecosystems and gamma-diversity (diversity in a region or country *etc.*). As per the breadth of the geographical area surveyed, biological diversity is measured in three ways. Alpha diversity is the number of species at one habitat in one locality. Beta diversity is the rate at which the species number increases as nearby habitats are added. Gamma diversity is the total number of species in all habitats across a broad area. For example: the number of species of *Saccharum munja* in the huge hectares of Babasaheb Bhimrao Ambedkar University Campus, Lucknow, India. If the survey is extended up to the South City Colony, Telibagh and Ashiana nearby, then the rate of increase in species count is beta diversity. If the extensive study is done in Lucknow, the species count is termed as Gamma diversity. Baseline data of species diversity is collected in terms of Index of species richness, species diversity and species evenness.

4.5 Taxonomic Classification of Biodiversity

Species is the most fundamental unit of measuring biodiversity. But the total biological diversity measurement require more in-depth and wider studies, wherein genera, families, orders, classes, phyla, kingdoms must be considered because the respective higher unit is assemblage of species in the lower units with a common ancestral linkage. This has lead taxonomists to develop the hierarchy of classification. Genus is the group of similar species with almost immediate common ancestry. Family is the group of genera kept together due to the basic similar characteristics identified and so on. Hierarchy of classification is given in the ascending order as follows:

Species

Genus

Family

Order

Class

Phylum

Kingdom

Whittaker has classified the biodiversity in five Kingdoms given as Plantae, Fungi, Animalia, Protista and Monera.

Measuring the biodiversity expanded within each level of the taxonomic hierarchy is important in terms of understanding evolutionary pathways, is an important feature of conservation steps. The construction of branching or successive splitting of species due to reproductive barrier, competition *etc.* to map evolutionary change is called cladistics and developing higher classifications (genus and further) accordingly is known as phylogenetic systematics.

Estimation of genetic diversity requires studies at the species or population level. Actual estimates of total genetic diversity are actually impossible. Some gene pools may be large (the maximum number of genes in a species is around 400,000) and continually changing whereas others are not.

The biodiversity can also be given category as exploited and underexploited depending upon the level of exploitation.

4.6 India as a Mega-diversity Nation or Biogeographic Zones of India

12 megadiversity countries exist in the world. India is one mega-diversity nation among those.

- ✰ Variety of climate (summer, winter, spring, monosoon).
- ✰ Variety of topography (Different regions: mountains, valleys, plataeus, plains, deserts).
- ✰ Variety of cultures, customs and religions.
- ✰ 49,219 (approx 50,000) plant species exist in India.
- ✰ 89,317 (approx 90,000) animal species exist in India.
- ✰ 167 crop species and over 350 species of wild relatives of cultivated crops exist in India.

1. Himalayas (Trans-Himalayas)

- ✰ It extends into Kumaon to Kashmir region.
- ✰ Major vegetation is forest including woody trees or timber trees or tree line.

  e.g. Sal: *Shorea robusta*, shisham : *Dalbergia sissoo*, oak tree, Babool (*Acacia sp.*) in dry areas.

  Small plants grow in the forest *e.g.* shurbs, herbs and grasses.

Inner Himalayas (Kashmir)

- **Trees**: Poplar trees (*Populus* sp.), woody trees
- **Agriculture**: Saffron (kesar), almonds, apple, peaches, walnut.
- **Animals**: Wild sheep, goats, cows, nilgai, deer, antelopes, butterflies, honey bees, bees, elephants, giraffe, birds: sparrow, crow, eagle, kite, nightingale, wood-pecker, pea-cock, cuckoo, vultures, owls, bats, mouse, rats, elephants, leopard, lion, bear, marbled cat, tortoise, turtles, squirrels, ants, mosquitoes, flies, termites, fire flies.

2. (i) Eastern Himalayas (Hotspot of Biodiversity in India)

- It includes Sikkim region and extends in the East. (Sikkim to Eastern) region.

Abiotic Conditions

- Higher rains
- Humidity and warmer abiotic conditions

Biodiversity

- Higher Biodiversity

Timber trees (woody trees or tree line): Shisham (*Dalbergia sissoo*), kachnar (*Bauhinia*), babool (*Acacia* sp.) in dry areas.

Flowering plants (flowers): Sunflowers, roses, lilies, marigolds, tulips, orchids, dahlias, seasonal flowers.

Birds: Sparrows, crow, peacocks, kites, vultures, cranes, owls, bats, eagles, nightingales, cuckoo, wood peckers, hen *etc.*

Reptiles: Snakes, crocodiles, lizards, turtles, tortoises.

Animals: Fox, bear, wild cat, rabbit, black bear, wolf, goat, deer, and antelope.

Insects: Butterflies, bees, honeybees, ants, catterpillars, earthworms, termities, mosquitoes, flies, fire flies.

3. Western Ghats (Hotspot of biodiversity in India)

- Near sea area (Maharashtra, Karnataka, Kerala, Tamil Nadu)
- Many plant species (>15,000)
- Trees *e.g.* Bamboo.

Endemic species of Western Ghats (also for endemism in India or endemic species in India)

Endemic plants of Western Ghats

- Dicotyledonous plants
- 245 species of orchids

Endemic animals of Western Ghats

- **Fish** (100 species of fish or many species of fish)
- **Birds** (16 endemic species of birds): Wynaad Laughing Thrush, Nilgiri Laughing Thrush.

4. Indo-Burma (Hotspot of biodiversity in India)

- North East of (Arunachal Pradesh, Assam, Andaman Islands).

Biotic Conditions

- High Biodiversity (Plants, Animals, Micro-organisms).
- Trees, Herbs, Shurbs, Grasses, mangrove vegetataion (Trees root respire by going up in air).
- Rabbit, Deer, Elephants, Fox, Wolf, Giraffe, Toads, Turtles, Lizards, Snakes, Peacocks, Eagles, Kukkoo, Sparrow, Crow, Crocodiles, fire flies *etc.*
- **Mammals (animals)**: Arunachal Macaque (*Mucaca munzala*)
  Laotian rock rat
  The leaf Muntjac
  The Bugun Liocichla
  The Khasi Hills Toad

5. Desert

- (Rajasthan + Kutch + Delhi + Gujarat) area
- **Climate** (Very less rainfall, < 70 cm)
- **Xerophytic plants**: *Acacia* (Babool), Prosopis, Cactus (*Opuntia*).
- **Animals** : Camel, Lizard, Fox, Wolf

6. The Ganga Plains (Gangetic Plains)

- (U.P. + Bihar+ Bengal) Plains
- Most fertile soil
- Rainfall: West U.P. (<70 cm or less rainfall) $\xrightarrow{\text{to}}$ Bengal (> 150 cm or more rainfall)
- **Vegetation**: Dry Decidous Forest $\xrightarrow{\text{to}}$ Tropical Moist Forest

  Trees

  Pipal (*Ficus religiosa*)
  Neem (*Azadirachta indica*)
  Mango (*Mangifera indica*)
  Arjun (*Terminalia arjuna*)
  Bargad (*Ficus* sp.)

Shisham (*Dalbergia sissoo*)
Dhak, Mahua, Tendu, Chirounji
Babool (*Acacia nilotica*) in dry areas

7. The Central India (The Semi-Arid Area)

- ☆ (M.P. + Orissa + Gujarat) Areas
- ☆ Woody Trees *e.g. Tectona grandis*
- ☆ Dry Areas (Babool) *Acacia* sp.

8. The Deccan

- ☆ Dry region (Andhra Pradesh + Tamil Nadu+ Karnataka)
- ☆ Rainfall (very less)
- ☆ Forests: Babool in dry areas
- ☆ Chandan/Sandalwood tree (*Santalum* spp.)

9. Andamans

- ☆ Islands near sea.
- ☆ Mangrove forests (Roots go up in air to respire through pneumatophores)
- ☆ Rice and sugarcane is grown. Arjun tree in dry areas.

10. Assam

- ☆ Highest rainfall (Cherapoonji)
- ☆ Insect eating plants, Grasses.
- ☆ **Woody trees**

  Pinus tree, *Shorea robusta* (Shisham), *Ficus religiosa* (Peepal).

4.7 Hotspots of Biodiversity

Very rich biodiversity in a defined area (is called as hotspot of biodiversity). *e.g.* variety of flowering plants, animals, insects, birds, reptiles, amphibians, butterflies. Most of the species found in the hotspot of biodiversity are endemic species (area specific species). The species are used for variety of purposes. Thus conservation steps are required (*in-situ* conservation).

- ☆ Norman Myers (1988) gave the concept of hotspot of biodiversity.
  - (i) Rich biodiversity in an area *e.g.* herbs, shrubs, grasses, trees, plants, animals, insects, birds, reptiles, amphibians, butterflies, insects *etc.*
  - (ii) Biodiversity is used for various uses (food, fibre, medicines, fruit, fodder, timber, beverages) and under *continuous threat (danger of extinction).*
  - (iii) The biodiversity is under continuous danger of extinction. They need conservation steps (*in-situ* conservation).
- ☆ **Important points of hotspots (key to designate a hotspot)**:
  - (i) High number of endemic species or higher degree of endemism *i.e.* Many species are found only in the particular area (area-specific species)

= endemic species (area specific species). The phenomenon of species restriction to a particular are is known as endemism. (***MTU, 2012-13, 2013-14***)

(ii) High degree of danger to species in a hotspot.

☆ 34 hotspots are identified in the world, 03 Hotspots are identified in India. Thus, India is a megadiversity nation.

Hotspot of Biodiversity in India

(i) Eastern Himalayas

(ii) Western Ghats

(iii) Indoburma

(1) Hotspots of Biodiversity in India (Eastern Himalayas, Western Ghats and Indoburma)

Definition of Hotspot

Areas with very rich biodiversity (variety of plants, animals, micro-organisms, insects *etc.*) are known as hotspots of biodiversity.

(i) The Eastern Himalayas

☆ It includes Sikkim region and extends in the East. (Sikkim to Eastern) region.

Abiotic Conditions

☆ Higher rains.

☆ Humidity and warmer abiotic conditions.

Biodiversity

☆ Higher Biodiversity.

Timber trees (woody trees or Tree Line): Shisham (*Dalbergia sissoo*), Kachnar (*Bauhinia*), Babool (*Acacia* sp.) in dry areas.

Flowering plants (flowers): Sunflowers, roses, lilies, marigolds, tulips, orchids, dahlias, seasonal flowers.

Birds: Sparrows, crow, peacocks, kites, vultures, cranes, owls, bats, eagles, nightingales, kukkoo, wood peckers, hen, *etc.*

Reptiles: Snakes, crocodiles, lizards, turtles, tortoises.

Animals: Fox, bear, wild cat, rabbit, black bear, wolf, goat, deer, and antelope.

Insects: Butterflies, insects, bees, honeybees, ants, caterpillars, earthworms, termites, mosquitoes, flies, fire flies.

Endemic Species of Eastern Himalayas

Endemic Plants of Eastern Himalayas

☆ 55 species of flowering plants are endemic and 'rare'. *E.g.* The pitcher plant (*Nepenthes khasiana*) (*Insect-eating plant or Insectivorous plant*).

Tree species

Ficus religiosa (Pipal), *Shorea robusta* (Shisham). *Piper longum* ("long").

Endemic Animals of Eastern Himalayas

- **Lizards and reptiles (turtles)**

  Over 60 per cent (many) reptiles and amphibians are endemic.

Endangered Species of Eastern Himalayas (of India)

- The Relict Dragonfly (*Epiophlebia laidlawi*).

Threatened Species of Eastern Himalayas (of India)

- One Horned Rhinoceros (*Rhinoceros unicornis*).
- Wild Asian water Buffalo.
- Mammals, Birds, Reptiles, Amphibians, Plant species.

2. Western Ghats

- Near sea area (Maharashtra, Karnataka, Kerala, Tamil Nadu).
- Many Plant species (>15,000).
- Trees *e.g.* Bamboo.

Endemic species of western ghats (also for Endemism in India or Endemic species in India) or Endemic plants of western ghats

- Dicotyledonous plants.
- 245 species of orchids.

Endemic Animals of Western Ghats

- **Fish** (100 species of fish or many species of fish).
- **Birds** (16 endemic species of birds)- Wynaad Laughing Thrush, Nilgiri Laughing Thrush.
- **Lizards and Reptiles (Turtles)**: Over 60 per cent (Many) reptiles and amphibians are endemic. Caecilians, Indotyphlus, Gegenophis, Uraetyphlus, Anurans, Toad Bufoides (Toads).
- Lion-tailed macaque (*Macaca silensus*).
- Nilgiri leaf monkey (*Trachypithecus johni*).
- Brown palm civet (*Paradoxurus).*
- Nilgiri tahr (*Hemitragus hylocrius*).

3. Indo-burma

- North East of (Arunachal Pradesh, Assam, Andaman Islands).

Biotic Conditions

- High Biodiversity (Plants, Animals, Micro-organisms).
- Trees, Herbs, Shurbs, Grasses, *mangrove vegetataion* (Trees root respire by going up in air).
- Rabbit, deer, elephats, fox, wolf, giraffe, toads, turtles, lizards, snakes, peacocks, eagles, kukkoo, sparrow, crow, crocodiles, fire flies *etc*.

☆ **Mammals (Animals)**: Arunachal Macaque (*Mucaca munzala*)
Laotian rock rat
The leaf Muntjac
The Bugun Liocichla
The Khasi Hills Toad

Economic Plants (Agriculture) of Western Ghats

Rice, banana, citrus, ginger, chilli, jute, sugarcane, 5 palms of commercial importance, coconut, arecanut, palmyra palm, sugar palm, wild date palm.

4.8 Threats to Biodiversity/How Man-made Extinction of Biodiversity is Caused? (*MTU, 2012*)

(i) Hunting

Animals are killed for enjoyment, to provide food, is known as hunting.

Biodiversity	*Use*
African Rhinos	Horns
Elephants	Ivory from tusks
Fishing	Fish

(ii) Poaching

Ill-legal trading of animals to obtain useful and costly materials, is known as poaching.

Biodiversity	*Use*
Chiru (*Pantholops hodgsoni*)	Very warm, light-weight, soft and expensive Shawl "**shahtoosh**" or "**king of wool**".

Chiru is killed to obtain the wool and now becoming extinct. It is a species of Tibet.

(iii) Habitat Loss or Habitat Fragmentation

Changes made in the habitat of organisms, cause loss of biodiversity. *e.g.* Trees are cut (**Deforestation**) for developmental activities (*i.e.* agriculture, mining, road development, construction). Animals lose their habitats. Habitat is the natural place where animals live.

Changes in Habitat ⟶ loss of their natural place to live

Trees are cut (Deforestation) ⟶ development (Agriculture, Mining, Construction) ⟶ animals lose their habitats

(iv) Introduction of Exotic (Foreign) Species

Plants and animals introduced in an ecosystem are known as exotic species. They are the new species introduced in an ecosystem. Thus, they grow and reproduce fast due to less environmental resistance *e.g.* No predators (animals who eat them) are there. Plants and animals introduced in an ecosystem are known as exotic species.

(v) Pollution

High temperature is caused by global warming and climate change. Polar Bear and penguins cannot survive at high temperature. They are under threat and becoming extinct.

Due to global warming, climate change

⇩

polar bears and penguins die in Antarctica

(vi) Pest Control System in Modern Agriculture

Excessive use of chemical pesticides kills the community of pests *e.g.* DDT kills pests.

(vii) Over-exploitation

Population is increasing exponentially. Increasing population puts pressure on Natural Resources (*i.e.* Trees, plants, animals, microorganisms, insects *etc.*) to fulfill the growing demands (for food, fibre, fodder, fruit, fuel, medicines, gums, resins *etc.*). It caused loss of biodiversity (**Figure 4.3**).

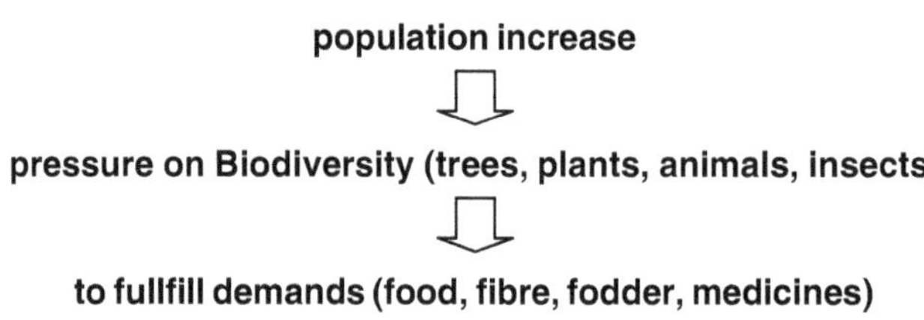

Figure 4.3: Loss of Biodiversity Due to its Overexploitation.

(viii) Diseases

Typhoid, malaria may cause death of humans and animals. Necrosis cause killing of tissues in plants, due to infection and air pollution.

(ix) Collection of Animals and Plants

Animals are kept in zoos and private collections for enjoyment, away from their natural habitat under human care. This reduces their ability to adapt in natural environmental conditions.

(x) Shifting and Jhoom Cultivation

Traditional farmers burn the crop residues and move to another place. The practice of burning the crop residues and shifting to another place is known as slash and burn agriculture. It also reduces the biodiversity, growing at the agricultural farm.

4.9 Endemic Species of India *(MTU, 2012)*

Many species are found only in the particular area (area-specific species), are known as endemic species (area-specific species). The phenomenon of species restriction to a particular are is known as endemism.

Endemic Plants of Eastern Himalayas

55 species of flowering plants are endemic and 'rare'. *e.g.* the pitcher plant (*Nepenthes khasiana*) (insect-eating plant or insectivorous plant).

Tree species: *Ficus religiosa* (pipal), *Shorea robusta* (Shisham).

Spices: *Piper longum* ("long")

Endemic Animals of Eastern Himalayas

- **Lizards and reptiles (turtles)**

  Over 60 per cent (Many) reptiles and amphibians are endemic.

Endemic Species of Western Ghats

Endemic Plants of Western Ghats

- Dicotyledonous plants.
- 245 species of orchids.

Endemic Animals of Western Ghats

- **Fish** (100 species of fish or many species of fish).
- **Birds** (16 endemic species of birds)- Wynaad Laughing Thrush, Nilgiri Laughing Thrush.
- **Lizards and reptiles (turtles)**

  Over 60 per cent (Many) reptiles and amphibians are endemic. Caecilians, Indotyphlus, Gegenophis, Uraetyphlus, Anurans, Toad Bufoides (Toads).
- Lion-tailed macaque (*Macaca silensus*).
- Nilgiri leaf monkey (*Trachypithecus johni*).
- Brown palm civet (*Paradoxurus*).
- Nilgiri tahr (*Hemitragus hylocrius*).

4.10 Endangered Species of India

- Liontailed macaque, Indian wolf, leopard, Indian lion, red panda, wild dog, Indian fox, jackal, barasingha, jungle cat, leopard cat, palla's cat, golden cat, marbled cat, great indian one-horned rhinoceros, wild buffalo, Assam rabbit, chinkara, wild buffalo, black deer, desert cat.
- **Reptiles**

  Indian Python, trunk turtle, green sea turtle.

- **Birds**

  Eastern Crane, blacknecked crane, hooded crane, great white crane, pinkheued duck.

Endangered Species of Eastern Himalayas (of India)

- The Relict Dragonfly (*Epiophlebia laidlawi*).

4.11 Threatened Species of India (Eastern Himalayas)

- One horned rhinoceros (*Rhinoceros unicornis*).
- Wild Asian water buffalo.
- Mammals, birds, reptiles, amphibians, plant species.

4.12 Conservation of Biodiversity or Biodiversity Management

Biological extinction has been a natural phenomenon in the geological history as a part of evolutionary process known as natural extinction. The rate of extinction was given as one species every 1000 years. Between 1600 and 1950, the rate of extinction increased upto one species every 10 years. Now it is extended to 1 species every year. The world's tropical rain forests are disappearing at an alarming rate and they are among the richest in terms of biodiversity. So they are highly exploited for the variety of usage from various bioresources. The tremendous progressive decline in the number of species of plants and animals lead to ecological crisis *i.e.* threatening the whole life-support system and the large number of habitats and species, leading to, the phenomenon of ecological disaster. Acceleration was observed in the rate of extinction of species as a result of human interference known as accelerated extinction. Urbanization, modern agriculture, construction, mining etc. are the examples of anthropogenic activities. Examples of extinct species from Indian fauna are Indian Cheetah, the pick headed duck and mountain quail. Besides these several species of mammals, 47 species of birds, 57 species of reptiles, 3 amphibians and a large number of butterflies, moths and beetles are facing danger of extinction. Similarly over 1500 species of vascular plants are considered to endanger. Developing countries of Europe and North America have depleted their forest resources, so limited the biodiversity.

Species are being continually lost due to lack of proper management. Modernization and consumerism associated with urban-industrial living style has caused an irreversible loss of bioresources. Poverty also compels the vast majority of people to use few accessible resources at their disposal at their fullest which is an unsustainable manner. UN Conference on Environment and Development has raised this issue in the global forum in Rio-de-Janeiro in June 1992 as: the relationships between unequal access to resources and the unsustainability of development and loss of biodiversity in particular. Technology intensive agriculture is progressively reducing variety in farms and also been the cause of environmental degradation.

Biodiversity management is an integrated approach towards its conservation to ensure sustainable development, whose operation depends on the feedback from the existing diverse ecosystem dynamics *i.e.* the pattern of energy flow, obtaining various resources and extracting energy pattern position of species, interaction among the species, pattern and causes of loss of variety of species as a result of natural phenomenon of extinction and anthropogenic causes like hunting and poaching for their products such as hides and skin, tusk, antlers fur, meant, perfumes, cosmetics, unecofriendly tourism, increasing population, pressure, urbanization and industrialization, pollution, destruction of habitat by man for its settlement, grazing grounds, agriculture, mining, industries, highway, construction, drainage habitat destruction (or alteration) and fragmentation, introduction of new species, war, climate change and industrial agricultural and forestry practices *etc*. Thus a conservation strategy involving a number of parameters such as number of species, their population dynamics, distribution, habitat, structure, microhabitat, physical environment, climate, present management and past historical scenario, is to be adopted in order to combat the global crisis of loss of biodiversity. The basic underlying principle is sustainable extraction of energy *i.e.* within the carrying capacity from the ecosystems. Therefore, the efforts for conservation of biodiversity should be in tune with the natural executing processes from micro level to mega level.

Finally, biodiversity conservation and use of genetic resources is a team work involving potential disciplinary contributions *i.e.* Plant breeding, biotechnology, taxonomy, agroecology, botany, anthropology, geology agroforestry, genetics, plant physiology, biochemistry, tissue culture, breeding, pedology, study of pollination and all the related disciplines. Thus the gene-pool must be interrelated to assess the diversity in totality. Therefore, taxonomists within the length and breadth of ecological range collectively analyze and assess the variety of data as a step towards its conservation.

Rich biodiversity has been observed near equator than near the poles, indicating the conservation priority being focused on the developing countries of the tropics. However, the global response being observed as a result of occurrence of pollution relates the cause and effects on each other. Thus, conservation scenario must be best described in both developing and developed countries perspective.

4.12.1 Types of Conservation of Biodiversity

(a) *In-situ* Conservation (in=inside), (situ=habitat or natural home) or Park Conservation

It is designated as comprehensive system of protected area. Conserving the plant and animal species within their natural habitat, is known as In-situ conservation. The forest areas are demarked as national parks, sanctuaries and biosphere reserves where biodiversity is conserved.

1. Consolidating the network of protected areas for wildlife to ensure the conservation of ecosystem and biogeographic units.
2. Establishing new protected areas based on utility and endangerment of species.
3. Ensuring conservation of biodiversity and rich ecosystems outside the network of protected areas.
4. Encouraging continuous and traditional agricultural practices.

5. Encouraging public participation in planning and management of protected areas.
6. Initiating regional co-operation for conservation of ecosystems and species.

(i) Protected Area

World conservation union defined "Area of land or sea for protection of biodiversity through legal or other effective means" *e.g.* National Parks, Sanctuaries, Biosphere Reserves.

National Park

Area of forest reserved (by government) for saving biodiversity, is known as National Park.

- ✰ Limited human activity (interference) is allowed in Buffer zone only.
- ✰ Particular wild life species is saved in national parks. *e.g.* Jim Corbett National Park and Dudwa National Park in Uttar Pradesh (for saving Tiger).

Sanctuary (Wildlife Sanctuary, Bird Sanctuary)

Area of forest reserved (by government) for saving wild animals and plants (biodiversity), is known as Sanctuary.

- ✰ Limited human activity (interference) is allowed.
- ✰ *e.g.* Timber harvesting, collection of minor forest products, private ownership rights.
- ✰ Bharatpur wildlife sanctuary in Rajasthan.

Biosphere Reserve

- ✰ Ecosystem oriented conservation approach
- ✰ All plants and animals species are conserved.
- ✰ Multiple land-use is permitted.
- ✰ No biotic interference is allowed except in buffer zone.
- ✰ *e.g.* Nanda-devi Biosphere Reserve (U.P.), Nilgiri Biosphere Reserve (Tamil Nadu, Kerela, Karnataka).

(b) *Ex-situ* Conservation (Ex=outside), (situ=habitat or natural home)

Efforts to save animals and plants species (endangered species) outside the natural habitats (under the human supervision). *e.g.* Botanical Gardens, seed-banks, gene-banks, zoos, aquarium.

Seed Bank

Seed of endangered plants are preserved in laboratory at low temperature (without damaging the quality of seed), is known as **cryopreservation**.

Gene Bank (Gene Pool or Gene Library)

Genes of endangered plants and animals are saved in laboratory to preserve the Genetic Diversity.

Zoos

Animals are kept in a space under human care.

Botanical Gardens

Plants are grown in artificial gardens under human care.

(2) *Ex-situ* Conservation

Sometimes the population of species may decline or may become extinct due to genetic or environmental factors such as genetic drift. In this case, a sample population is conserved in inbuilt infrastructure.

Advantages of *Ex-Situ* Conservation

(i) Food, Shelter, Security is provided to endanger animals.

(ii) Increase of life-span of endangered animals.

(iii) Increasing the chance of reproduction.

Disadvantages of *Ex-situ* Conservation

(i) Ability of plant and animal species to adapt to natural environment decreases.

(ii) Only few species (Less number of species) are saved.

4.12.2 Conservation Tools (Techniques) of Biodiversity

(i) Genetic Engineering

Genes of desired characters are introduced in species to develop resistance.

(ii) GIS (Geographic Information System)

Remote sensing helps in identifying areas of biodiversity threat and extent of loss of biodiversity.

4.13 International Efforts to Save Biodiversity

(i) IUCN (International Union for Conservation of Nature and Natural Resources), Headquarter

Switzerland is an international organization which maintains **red-data book** in which list of endangered species of plants and animals are given.

Categories of Biodiversity

(a) Extinct: Species of plants and animals which are finished in number are known as extinct. Extinct species of India are Dodo and Passenger Pigeon.

(b) Endangered: Number of species has been reduced to a certain extent that if such species is not protected, they are in immediate danger of extinction, are known as endangered species. Endangered species of India are red panda, tiger, lion, orchids, indian peacock, *Piper longum etc.*

(c) **Vulnerable:** Number of species is still abundant but under a serious threat of becoming endangered are known as vulnerable 'or' number of species if not saved, then may come under the category of 'endangered'.

(d) **Rare:** Species which are not endangered or vulnerable, but are at risk are known as Rare.

(e) **Threatened:** Number of species is less, if not saved, they may become extinct, are known as threatened.

Threatened= Endangered+ Vulnerable+ Rare.

Indeterminate species: Species which are less in number, but the reason of their scarcity is not known are known as indeterminate species *e.g.* Snow leopard.

Endangered Economic Plants

Gymnosperms- (*Pinus* sp.), Sandalwood tree (*Santalum album*)

Endangered Ornamental Plants

Rhododendron sp., Orchids

Endangered Medicinal Plants

Atropa, Rauwolfia serpentia

Endangered Animals of India

Red panda (*Ailurus fulgens*)

(ii) The Biodiversity Treaty or (CBD) "Convention on Biological Diversity"

The Convention on Biological Diversity (CBD) was signed by participating countries at the 1992 United Nations Conference on Environment and Development at Rio de Janerio. The Convention was ratified by the New Zealand government, in September 1993, making it a legally binding document.

The CBD establishes a framework at local, national and international levels to conserve and sustainably utilise biological diversity. It has three major objectives, described as:

- The conservation and values of biological diversity.
- The sustainable use of biological resources.
- The fair and equitable sharing of the benefits arising out of the utilisation of genetic resources.
- Promotion of traditional knowledges and communities.
- Promotion of Biotechnology and Genetic Engineering to save genetic resources.

(iii) Ramsar Convention on Wetlands

An international conference to save wetlands.

4.14 Values (or uses or importance or benefits) of Biodiversity

Biodiversity has high practical value in terms of providing immediate economic gains of crops, fuel wood, fibre, food, medicine, natural pesticides, industrial chemicals, timber and non-timber products and ecological services including water and air purification, natural pollination, soil formation and maintenance, pest protection for crops and livestock, ground water recharge, watershed protection, buffering of floods and droughts and sequestration of carbon.

Biodiversity is an important natural resource. (*i.e.* present in nature and useful to man).

(i) It provides us various useful products. (*e.g.* food, medicine, wood, crops, fruits, wool, industries, fibres).

(ii) Ecological processes are fulfilled by biodiversity (*e.g.* air purification, water-cycle, green-plants carry out photosynthesis, food-chain and food-web).

(iii) Biodiversity helps maintain genetic diversity. Variations in the genes in the individuals of a single species, help adapt to environmental stress.

Thus living organisms (plants, animals and micro-organisms) are exploited to get the variety of uses. Excessive exploitation of biodiversity puts it into threat or danger of extinction. Thus there is a need to conserve or save Biodiversity.

The peculiarity of biodiversity is that it is a biological resource meant to provide: the material basis for human life, cultural and utilarian needs, nutritionally significant edible plants and animal species, medicinal drugs for human and animal health care, support to fisheries through coral reefs, place for religious practices and aesthetic beauty to the landform. Fungus *Penicillin* produced penicillin, the bark of the yew tree offers a treatment for various types of cancer, and an Asian viper's venom is used in a stroke-prevention medicine. Chemicals from sea sponges and some marine organisms might block arthritis inflammation and fight cancer.

Agar-agar which is extracted from the algae: *Gelidium* sp., *Gracilaria* sp. and *Hypnea* sp. is used in food industries as thickener and stabilizer. In pharmaceutical and microbiological research, it is used as culture media of microorganisms and plants. It is also used in creams, lotions, soaps and ointment. Sea cucumbers or holothurians are widely distributed in marine environments, from intertidal zones to the deep-sea bed. They can easily be seen lying on the sandy seabed, among corals or rocks. It is an important food for the Chinese, regarded as tonic food, and occasionally used as medicine. It is considered principally a banquet food in Taiwan whereas much as 95 per cent of sea cucumber is consumed at banquets and in restaurants. One of the most commonly sold freshwater turtle species in peninsular Malaysia is the Asian Box Turtle *(Coura ambainensis)* which is harvested for food. They are commonly found in rice fields, swampy areas and streams in rubber and oil palm Plantation. Seaweeds are commercially important living marine sources that belong to the primitive group of non-flowering plants (Thalophyta). They grow submerged in intertidal, shallow and sometimes subsurface water up to 100 m deep in the sea and also in brackish water estuaries. Wildlife is primarily used in Vietnam as a source of food.

Various biological organisms like plants, animals, insects and microorganisms help benefit agriculture, which forms the basis of many innovative traditional techniques. Farmers utilize these as alternatives to synthetic chemicals for pest and predator control.

Local people get benefited by social gains in terms of livelihood security, social satisfaction and aesthetic valuation. Natural Beauty is inherent, and not in the eyes of beholder. The beauty of individual organisms, landscapes and organic processes marks the transformative influence of aesthetic fulfillment. Our cultural identity is deeply rooted in our biological environment. Moral and religious values are attached to most of the plants and animals. Various festivals of local tribes are dependent upon the ancient plants and animals *e.g.* *'Amiah'* is festival of mangoes among the forest tribes. Various plants and animals are worshipped as God and Goddesses *e.g.* Snake: 'Nag-devata', Cow: 'Gao devi', Pipal tree, Bargad tree *etc.* Even various Goddesses are reported to be seated on their respective animals ('vahan') *e.g.* Goddess 'Saraswati' is sitted on 'Swan'. This actually indicates the prime importance of biodiversity. So the importance of biodiversity to the local people can be well understood. Primary biodiversity conservation movements in history have been started by the local community *viz.* Chipko movement started by Bishnoi community under the active leadership of Smt. Amrita Devi. People hugged the trees against the protest of cutting them. Later on the movement gained international importance under the guidance of Shri SunderLal Bhaguna in Tehri Garwal.

Global biodiversity strategy described by the sponsorship of World Resource Institute gives the philosophy as biodiversity reinforces economic and social security, which will in turn reinforce biodiversity. The total value of biodiversity cannot be easily estimated, due to lack of information and uncertainty and variation from location to location *i.e.* country to country and region-to-region.

The advancement in science and technology reduces the importance of biodiversity. Biotechnological progress almost nullifies the actual dependence on bioresources, thus devalue them. With the techniques like genetic engineering, tissue culture techniques a useful biological product can be synthesized chemically, the peculiarity of the product yielding respective species is definitely reduced. At the same time, the crop breeding industry develops the crops with highly desirable traits. There are many successful examples of use of wild genes to improve domestic crops.

Floral biodiversity also has tremendous potential for medicinal values. Plant extracts are widely used in drugs. Bio pesticides have now been extensively used in agriculture as a substitute of chemically resistant pesticides. The calabra been found in West Africa contributed to the development of methyl carbamate insecticides. The roots of a South American forest vine used by Amazonian Indians to stun fish became a source of a widely used biodegradable pesticide called rotenone. Vanoshadhi Chandrodaya is a renowned Indian book would be better described as bible of medicinal plants. Some of the important medicinal plants desribed are Arjun tree (*Terminalia arjuna*), Aloe vera, Neem and Tulsi.

(a) Direct Values

Consumptive Values and Productive Values

We get useful products from biodiversity. (Fruits, vegetables, wide array of food products, beverages, medicines, fibres, other products *e.g.* wood). Aloe vera is a source of nutrients (vitamins: A,B,C,E,M; minerals: Calcium, phosphorus, iron, magnesium, potassium, sodium; 22 amino acids: essential amino acids: isoleucine, leucine, valine, lysine, methionine, phenylalanine, threonine; enzymes: phosphatase amylase, lipase, cellulose, catalase; lignin, saponins and anthraquinones).

(b) Indirect Values

(i) Non-Consumptive Use Value

Values that are having important role for providing suitable conditions for living organisms are known as non-consumptive values.

Ecological Values

Following services are provided by biodiversity *e.g.*

- ☆ Soil formation
- ☆ Air purification
- ☆ Water cycle
- ☆ Carbon-fixing through photosynthesis
- ☆ Pollination
- ☆ Habitat to many animals

Medicinal Value (Table 4.1)

Table 4.1: Medicinal Values of different Herbs.

Medicine or Medicinal Value	*Biodiversity (from which medicine is obtained)*
Quinone	*Chinchona* sp.
Blood purifier	*Terminalia arjuna*
Antipyretic (to get rid of fever)	*Ocimum* sp. (Tulsi)

Detoxification Mechanism of Aloe vera

- ☆ Toxicants are absorbed in the villi of intestine.
- ☆ *Aloe barbadensis miller* absorbs these toxicants, removes from digestive system, which aids in health and provides rich source of nutrients.

Pest Control Values

Biopesticides help kill pests *e.g.* Neem.

(ii) Aesthetic, Social, Cultural Value

Many festivals are based on worship of biodiversity *e.g.* Nag-panchami, Basant panchami. Hindus worship Peepal tree (air pollution tolerant tree), cow as mother *etc.*

(iii) Recreational Values

Biodiversity is useful in providing following enjoyment activities: fishing, camping, night safari, tracking, bird watching, photography, site-seeing, horse-riding, camel-riding *etc.*

Participatory Learning

(a) Objective Questions

1. Variety of living organisms is known as ______.
2. Saving wildlife in their natural habitat is known as ______.
3. Illegal hunting of wild animals is known as ______.

(b) Very Short Answer Type Questions

1. Define biodiversity.
2. What is the need for biodiversity conservation?
3. Define endangered, endemic, threatened and vulnerable species.
4. Define biomes.
5. What kind of vegetation is found near coastal areas or wetlands? Or what kind of adaptation plants do to survive in high salt stress? Or what are pneumatophores? Or what is mangrove vegetation?
6. What are insectivorous plants? Give example. Why they do so?
7. Define endemism and endemic species: Name Two. (***MTU, 2013-14***)

(c) Short Answer Type Questions

1. Write short-notes on
   (a) IUCN (b) IUCN categories of wild-life
   (c) CITES (d) TRAFFIC.
2. What are the hotspots of biodiversity?
3. What are the measurements of biodiversity?
4. What are the causes of extinctions (***MTU, 2012***).
5. List the medicinal plants with their respective properties.
6. Differentiate between hunting and poaching.
7. What is the role if IT in biodiversity conservation.

8. Explain plant tissue culture technique.

9. Explain the role of biotechnology in biodiversity conservation.

(d) Long Answer Type Questions

1. Comment " India as a mega diversity nation." Or Explain biogeographical classification of biodiversity. (***UPTU, 2006***)

2. What do you understand by hotspots of biodiversity. What are the salient features? (***UPTU, 2005***)

3. What are the different categories of biodiversity? Explain their values? (***MTU, Carryover 2013***)

4. What do you understand by biodiversity conservation or management? What are the methods of biodiversity conservation?

5. How genetic diversity originates? Discuss survival of fittest with respect to genetic diversity and Darwinian theory of existence.

Answers to Objective Questions

1. Biodiversity
2. *In-situ* conservation
3. Poaching

Chapter 5

Energy Resources

5.0 Daily Life Examples: Save Energy Cartoon Series

Papa:	*Energy is ability to do work (by machines and by living things) as u know in Physics. Energy can be changed from one form to another but it can neither be created nor be destroyed.*
AKSHAY:	*Wow! That means moving a car needs energy. Lighting a lamp also consumes energy, Mummy cooks food with the help of heat energy!!!!!* *Where does this energy comes from?*
Papa:	*Car gets energy by burning fuel e.g. Coal, Petroleum, CNG Lamp gets electric energy, Cooking also consumes LPG.*

5.1 Population Explosion (MTU 2010, 2011)

Population is increasing exponentially (*i.e.* at a very faster rate), puts pressure on natural resources (*e.g.* air, water, soil, and other energy resources), to fulfill the growing demands of population *i.e.* food, clothing and shelter. The developmental activities also consume our important natural resources. Non-Renewable energy resources take a very long passage of time in their formation, so the pressure on non-renewable energy resources increases to a greater extent (**Figure 5.1**). Keeping in view

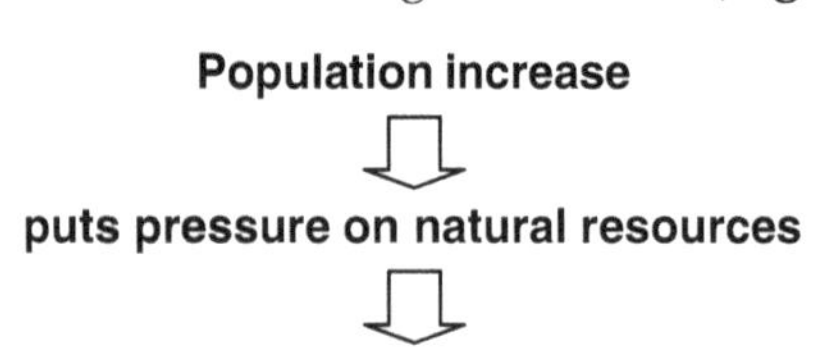

natural resource crisis **or** natural resource crunch or depletion of natural resources
e.g. crisis of fossil fuels (CNG, LPG, Petroleum), energy crisis.

Figure 5.1: Energy Crisis in Present Scenario.

Table 5.1: Difference between Perpetual and Renewable Energy Resources.

Perpetual (Inexhaustible) Energy Resources				*Renewable Energy Resources*		
Solar Energy	*Wind Energy*	*Tidal Energy*	*Hydro Energy*	*Geothermal Energy*	*Hydrogen Energy*	*Biomass Energy (Biomass=Organic Waste)*
(i) Direct heating (Radiant energy) *e.g.* solar cooker. (ii) Photovoltaics *e.g.* solar cell. (iii) Passive (iv) Active	Wind falls on Blades of windmill rotates to produce electricity.	Tides waves of sea rotate turbine, to produce electricity.	Water falling from height (kinetic energy) rotates turbine, to produce electricity. (Also known as hydro-electricity or hydro-power or river-valley projects or dam-benefits and problems)	Heat present inside the Earth, is used to rotate turbine, to produce electricity.	Fuel cell burns H_2 to produce electricity. $H_2 + O_2 \rightarrow H_2O + 2e$ (electricity)	(i) Fuel (Direct burning of biomass *e.g.* wood, paper, crop residues)
Advantages						
(i) Available in large amount. (ii) Eco-friendly or pollution free. (iii) Cheaper.	(i) Available in large amount. (ii) Eco-friendly or pollution free. (iii) Cost of production is not high. (iv) Energy can be obtained in large open areas.	(i) Available in large amount. (ii) Eco-friendly or pollution free. (iii) Energy can be obtained in large open areas.	(i) Available in large amount. (ii) Eco-friendly or pollution free. (iii) Cheaper than Thermal Power Plants. (iv) Other benefits: Fishing, Dams increase the beauty of an area.	(i) Energy can be extracted continuously. (ii) Eco-friendly or pollution free. (iii) No major land area is required. (iv) Additional units of geothermal power plants can be installed if required.	Hydrogen as a fuel (i) Very high calorific value. (ii) Ecofreindly.	(i) Organic waste is available in large amount. (ii) Ecofreindly. (iii) Cheaper (iv) India is an agriculture based country. Waste of biogas is manure.

Contd...

Table 5.1–*Contd...*

Perpetual (Inexhaustible) Energy Resources				*Renewable Energy Resources*		
Solar Energy	*Wind Energy*	*Tidal Energy*	*Hydro Energy*	*Geothermal Energy*	*Hydrogen Energy*	*Biomass Energy (Biomass=Organic Waste)*
Disadvantages						
(i) Availability depends on climate. (ii) Initial cost of installation is high.	(i) Availability depends on climate. (ii) Noise pollution due to rotation of blades and turbine. (iii) Deforestation	(i) Availability depends on climate (occurrences of tides). (ii) Cost is high.	(i) Deforestation. (ii) RIS (Reservoir Induced Seismicity): Probability of earth quakes increases. (iii) Cost is high. (iv) Water-borne and Water-induced diseases. (v) Displacement of many people.	(i) Less production efficiency. (ii) Earthquakes (iii) Air Pollution-H_2S (iv) High cost. (v) Noise pollution in drilling.	(i) Less availability of hydrogen in pure form. (ii) Difficulty in handling, storing and transporting.	(i) Biogas should be used within 10 m. of gobar gas plant.

the present energy crisis, government is promoting the use of renewable energy resources to reduce pressure on non-renewable energy resources. The concept to promote the use of renewable energy resources is known as **green energy concept**.

- ☆ So the natural resources are getting less in availability.
- ☆ (Scarcity of natural resource) causing the situation of population–explosion (exponential rise in population + scarcity of natural resources)

5.2 Classification of Energy Resources

(i) Renewable Energy Resources (Non-conventional energy resources)

The sources of energy which are renewed again and again at a faster rate in nature are known as renewable energy sources. *e.g.* solar, wind, hydro, geothermal, ocean, hydrogen, biomass.

(ii) Perpetual Energy Resources (In-exhaustible energy resource)

Among renewable energy resources, certain energy resources are present in nature in large amount are known as perpetual energy resources and are inexhaustible in nature. *e.g.* solar, wind, hydro.

(iii) Non-Renewable Energy Resources (Conventional energy resources)

These energy resources are formed inside the earth crust at very high temperature and pressure, takes a very long passage of time in their formation. Thus they are present in limited amount in nature's reservoir. Non-renewable energy resources are finite and exhaustible. *e.g.* coal, petroleum, natural gas, timber, nuclear fuels (**Figure 5.2**).

5.3 Major Differences between Conventional and Non-conventional Energy Resources

Table 5.2: Differences between renewable and non-renewable energy resources

Conventional Energy Resources	*Non-Conventional Energy Resources*
Conventional energy sources are **Non-Renewable** energy resources.	Non-Conventional energy sources are **Renewable** energy resources
Technologies are developed to exploit the potential of conventional energy resources	Technology development is under progress.
They are finite resource of energy.	They are available in large amount.
Combustion of fossil fuels releases lots of toxic gases in the atmosphere	They are the ecofriendly source of energy.
Examples: fossil fuels, nuclear fuels	Examples: solar, wind, hydro-electricity, geothermal, tidal *etc.*

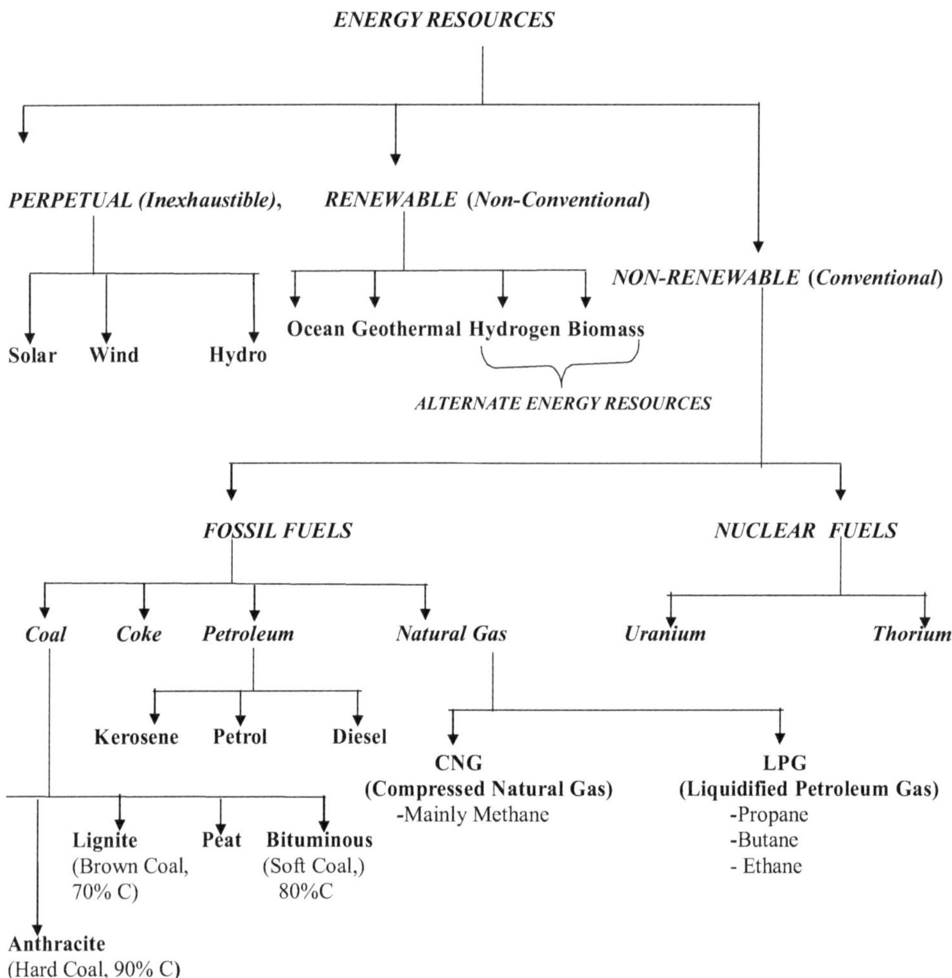

Figure 5.2: Classification of energy resources

5.3.1 Renewable (Non-conventional) Energy Resources

The energy resources are formed again and again by natural processes at a faster rate. They are used indefinitely. *e.g.* solar energy, hydropower, wind energy, geothermal, ocean energy. The major limitation in promoting the renewable energy is its high cost of production. Government is promoting research to develop applications of renewable energy resources.

Renewable energy resources are the major future energy resources *i.e. renewable energy future.* Many countries have already started using renewable energy resources at a commercial scale. Even in our country India, many commercial buildings are running their centralized air conditioning by the help of solar panels *e.g.* Tata Energy

Research Institute (TERI) at Lodhi road, New Delhi. Brazil has started using ethanol, produced from fermentation of sugarcane as a fuel to run cars and trucks. California has started using wind and geothermal energy. Middle east countries have realized to utilize the potential of solar energy substantially.

5.3.1.1 Solar Energy Applications

Solar energy is available in very large amount, thus considered perpetual energy resource and is considered major future source of energy.

Ultimate source of energy in an ecosystem is Sun. Solar energy incidents on green plants. They carry out Photosynthesis in the presence of CO_2, water and sunlight and make their own food, thus maintain life. Their death produces biomass. Solar energy influences earth's climate. Wind energy, biomass and hydropower are the resultant of solar energy. Wind blows from areas of low temperature to high temperature. Rainfall is the resultant of cloud formation due to evaporation of water in the presence of sunlight.

(1) Solar Energy

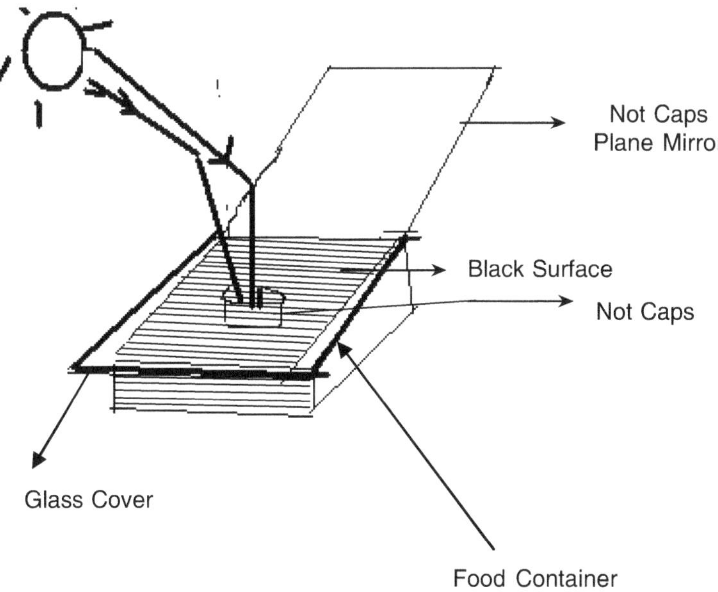

Figure 5.3: Diagram of Solar Cooker.

Photovoltaics

Sun ⟶ Silicon, Germanium ⟶ Electricity Production

Figure 5.4: Electricity flow in semiconductors by photovoltaics

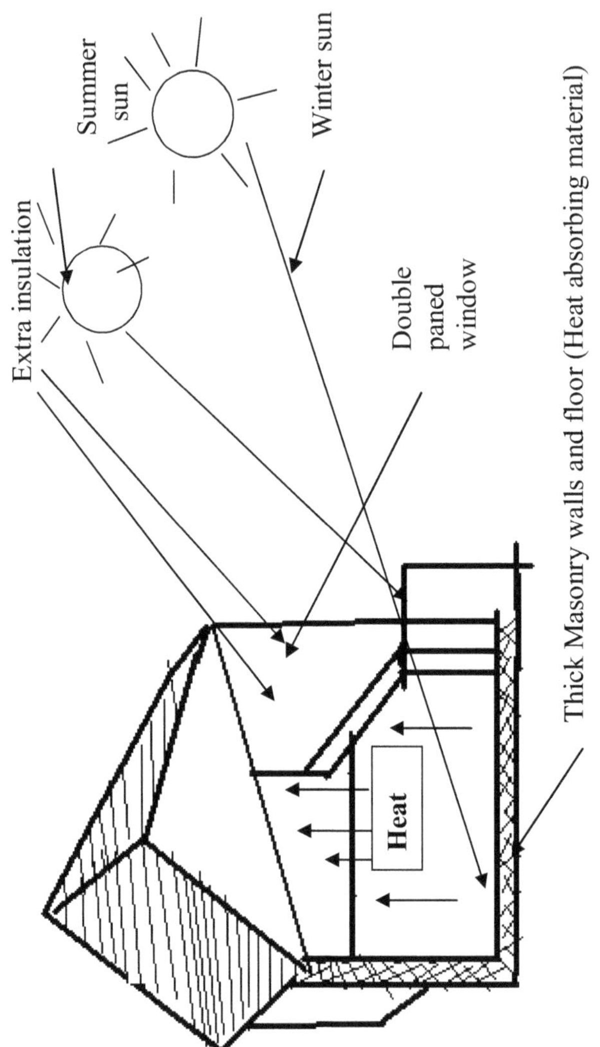

Figure 5.5: Solar Passive Heating.

Photovoltaics: Conversion of Solar Energy to Electricity

Semiconductors when irradiated with sunlight, electrons present in their ground state get energized to jump to an excited state *i.e.* flow of electrons occur and generates electricity. The phenomenon is known as Photovoltaics (PVs). Thus, Photovoltaics are the arrays of cells consisting of semiconductors that effectively converts solar radiation into electricity (Direct Current), used to power bulb or equipment or to recharge a battery. An inverter converts DC to AC for grid connected power generation (**Figure 43**).

Flow of Electrons in a Semiconductor when Irradiated with Sunlight

When n-type and p-type semiconductors put together and irradiated with sunlight, the excess electrons in the n-type material flow to the p-type and this way, movement of holes start. Thus electric field is generated at the junction of the two semi-conductors (known as p-n junction).

Other Applications of Solar Energy

- ☆ Solar cooker
- ☆ Solar water heater
- ☆ Solar furnace
- ☆ Solar power plant
- ☆ Solar toys
- ☆ Solar air-conditioning
- ☆ Solar vehicles
- ☆ Solar chimney
- ☆ Space cooling and heating
- ☆ Day lighting (solar street lights and traffic lights)
- ☆ Solar desalination
- ☆ Solar electricity- Photovoltaics
- ☆ Solar electricity-Thermal

Passive Solar House

Buildings can be designed in such a way to use active technology of solar space heating. It helps capture solar energy to provide space heat (**Figure 5.5**).

Passive solar design technique is brought about with following features:

(i) Sunlight passes through south-facing windows.

(ii) Interior walls, floors of brick, tiles or cement (made up of heat absorbing material) help absorb incoming sunlight.

Advantages of Solar Energy

(i) Renewable energy resource: Solar energy is available in very large amount.

(ii) Ecofreindly: Solar energy is *pollution-free i.e.* does not cause air pollution during its use like traditional energy sources.

(iii) **User friendly:** Solar powered instruments are easy to install and use. *e.g.* solar panels, solar water heaters, solar lighting, solar pumps, solar fountains.

(iv) **Long life and low maintenance:** Solar panels are required to be kept dust snow free for allowing maximum contact with sunlight. Life of solar panels is longer (approx. 25 years) and requires less maintenance.

(v) **Cheaper:** Solar system based electricity is a cheaper source of energy as compared to energy obtained from conventional electric systems.

(vi) **Reducing pressure on non-renewable energy resources:** Since solar energy is available in very large amount, its use reduces the pressure on non-renewable energy resources, which takes a very long passage of time in their formation, once consumed. Thereby, using solar energy and other renewable energy resources help save nature's reservoir of non-renewable resource, helps in *sustainable development*. It is a step towards sustainable future.

(vii) **Solar Photovoltaics** are used for the electricity generation in remote and isolated areas *e.g.* forest, hills, deserts. PV is frequently used in watches, pocket calculators and toys.

Disadvantages of Solar Energy

(i) **Availability of Sun**: Solar energy production is hampered when the Sun is not available (*i.e.* during nights, less sunny days and cloudy days).

(ii) **Sound technology:** Technology is to be kept advancing in such a way to efficiently capture the solar radiations over a larger area.

(iii) **Cost of installation:** Initial cost of installation is high. In case of photovoltaic cell, price of silicon wafers makes it very costly.

(iv) **Longer duration in solar cooking:** Solar cooking is a long time taking process. The food kept in the container can be over-heated if not removed from solar-cooker timely.

(v) **Variety in cooking:** Solar cooker cannot work as a substitute of LPG chullas because all type of foods cannot be prepared in solar cooker. *e.g.* chapatti and puri.

5.3.1.2 Hydro Energy

Water is made to fall from height to rotate turbine, which help generate electricity, is known as Hydro Energy.

Advantages

(i) Clean source of energy (eco-freindly or non-polluting).

(ii) Cheaper than thermal power plants.

(iii) Other benefits *e.g.* fishing, beauty of land area.

Disadvantages

(i) **Reservoir Induced Seismicity (RIS)** The continuous pressure of water column on the earth core increases the probability of occurrences of earthquakes.

(ii) Water-borne and water induced diseases are caused.

(iii) Flood is caused if leakage occurs in the dam.

5.3.1.3 Wind Energy

Wind is allowed to rotate the blades of windmill, which rotates turbine to generate electricity.

Advantages

(i) Cost of production is not high, can be reduced by research.

(ii) Initial investment is moderate.

(iii) Operation and maintenance cost is not high.

(iv) Eco-friendly.

(v) Available in large amount.

(vi) Produce more energy due to high efficiency.

(vii) Land below turbines can be used for growing crops and stock (animal) grazing.

Disadvantages

(i) Availability depends on steady (continuous) winds or climate.

(ii) Unsteady winds affect power production.

(iii) Noise pollution (a) due to rotation of blades; (b) due to rotation of turbines.

(iv) Large open area is required for electricity production.

5.3.1.4 Marine Energy or Ocean Energy

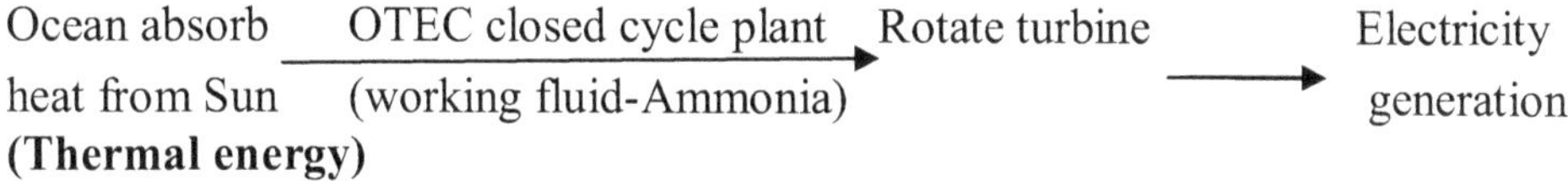

Figure 5.6: Production of electricity with the help of ocean energy

Advantages

(i) Available in large amount.

(ii) Eco-friendly.

(iii) Energy can be obtained near sea.

Disadvantages

(i) Installation of ocean thermal energy conversion (OTEC) power plant is costly.

(ii) Highly equipped engineers are required.

5.3.1.5 Tidal Energy

Tides of sea waves help rotate turbine to generate electricity, is known as tidal energy.

Advantages

(i) Available in large amount

(ii) Eco-friendly.

(iii) Energy can be obtained near sea.

Disadvantages

(i) Availability depends on steady (continuous) tides.

(ii) Unsteady winds affect power production.

(iii) Noise pollution (a) due to rotation of blades; (b) due to rotation of turbines.

(iv) Large open area is required for electricity production

5.3.1.6 Geothermal Energy

Hotwater or springs on the soil surface, is made to rotate turbine, to produce electricity, is known as geothermal energy **Figure 5.7**.

Advantages

(i) Geothermal energy can be continuously extracted.

(ii) It is an eco-friendly source of energy.

(iii) No major land area is required.

(iv) Installation of additional energy production units is possible, if needed.

Disadvantages

(i) Production of geothermal energy causes air pollution due to release of H_2S gas.

(ii) Earthquakes may result in the region of exploitation of geothermal energy, unless cooled water is not injected back into the reservoir.

(iii) Production of geothermal energy require large cost.

(iv) High safety measures are to adopted in geothermal power plant.

(v) Drilling operations in the geothermal reservoir causes noise pollution.

5.3.2 Alternate Energy Sources

5.3.2.1 Hydrogen as a Fuel for Future or Hydrogen as an Alternative Energy Source (*UPTU, 2009, 2010, MTU, 2011*)

Hydrogen is burned electrochemically with oxygen to produce electricity. Electrodes are dipped in electrolyte preferably Sodium Hydroxide. Hydrogen and oxygen is passed in the fuel cell and following chemical reaction generates electricity:

$$H_2 + O_2 \longrightarrow H_2O + \text{electrons}$$

(flow of electrons is electricity)

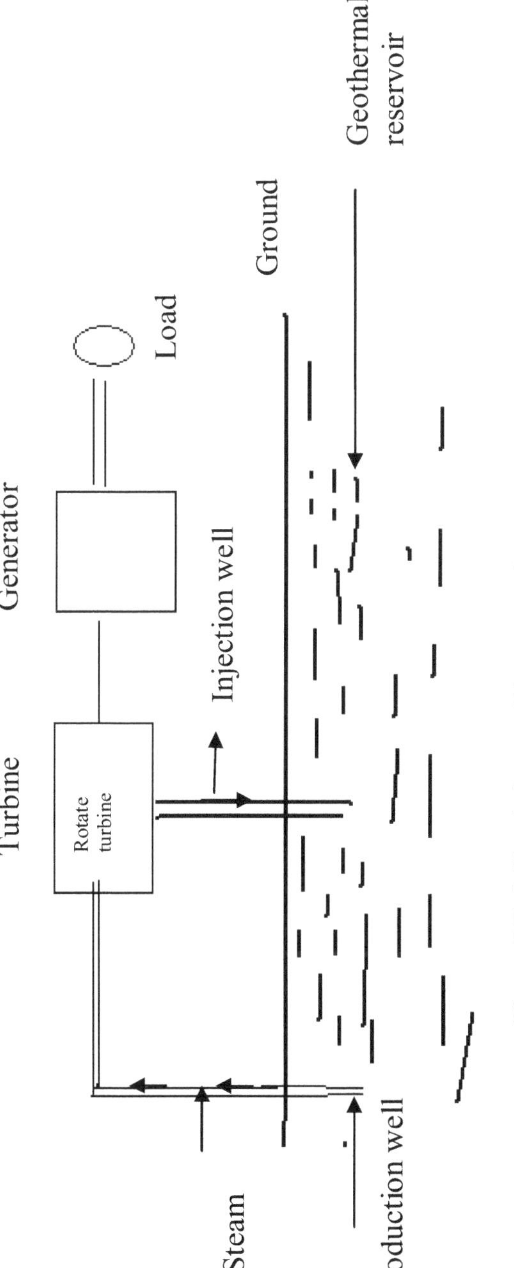

Figure 5.7: A Simple Layout Diagram of Geothermal Power Plant.

Advantages of Hydrogen

(i) Hydrogen has highest calorific value (150 kj/g). Thus it is an excellent fuel.

(ii) It is an eco-freindly source of energy.

(iii) Source of hydrogen (water) is present in large amount.

(iv) Simplest and lightest fuel in gaseous form.

Disadvantages of Hydrogen

(i) Hydrogen gas is highly explosive (inflammable).

(ii) High safety measures are required in storage and transportation.

(iii) Production of hydrogen is costly.

Production of Hydrogen

Large scale production of hydrogen require very large amount of energy. Its production becomes possible by thermal dissociation, photolysis or electrolysis or hydrolysis of water (or sea water).

5.3.2.2 Microbial Fuel Cell

Electrical Generation Process

When micro-organisms consume a substrate such as sugar in aerobic conditions they produce carbon dioxide and water. In case of absence of oxygen, microorganisms consume sugar and water to produce carbon dioxide, protons and electrons as described below:

$$\underset{\text{(sugar)}}{C_{12}H_{22}O_{11}} + 13H_2O \longrightarrow 12CO_2 + 48H^+ + 48e^- \quad (\textit{Eqt. 5.1})$$

Microbial fuel cells use inorganic mediators to tap into the electron transport chain of cells and channel electrons produced.

5.3.2.3 Energy Context with Respect to Indian Scenario (*UPTU, 2008, 2010, MTU 2011*)

India is an agriculture based country. Agriculture and dairy farming produces large amount of biomass *e.g.* agricultural waste (crop residues) and cowdung (gobar). Dead organic matter or dead plant and animal parts are known as biomass *e.g.* wood, leaf parts, and crop residue. The major crop residues were paddy straw, wheat straw, jowar straw, sugarcane trash, groundnut holms, maize stalks and bajra straw, accounted for almost 88 per cent of the total residues in the country.

(i) Directly as Fuel

Burning of biomass in 'chulhas' in villages is an example of direct use as fuel. The burning of dung (gobar) destroys essential nutrients of soil *viz.* N (Nitrogen) and P (Phosphorous). It is therefore, more useful to convert biomass into biogas and biofuels.

(ii) Biogas (Gobar Gas)

Methanogenic bacteria undergoes anaerobic decomposition of organic waste at high temperature (38-45°C) (thermophillic bacteria) under the ground known as digestion well or anaerobic digestor (AD) (**Figure 5.8**).

(a) **Physical and chemical properties of biogas:** Biogas burns with a blue flame and has high specific heat.

(b) **Production of biogas:** Biogas is produced by anaerobic decomposition of organic waste at 38-45°C (high temperature).

(c) **Chemical composition:** mainly methane (CH_4). Biogas is mixture of following gases: (a) CH_4 (40 per cent), (b) CO_2, (c) H_2, (d) N_2, mainly methane.

(d) **Waste material of biogas:** The semi-solid waste of biogas production is sludge which is highly nutrient rich and used as manure.

Advantages of Biogas

(i) It is an eco-friendly clean fuel.

(ii) Without storage tank, it can be supplied directly to the homes from plant.

(iii) Pathogens and parasites cannot come in contact of fecal material as the digestion of waste takes place in closed chamber.

(iv) Waste of Biogas (sludge) is used as highly nutrient rich manure.

Disadvantages of Biogas

(i) It can be supplied only to few kilometers of production of biogas.

(ii) **Biofuels:** Biomass is processed to generate ecofriendly fuel known as biofuels. India, bein an agriculture based country, possess great potential in biofuel industry **(Table 5.3)**.

Advantages of Biofuel

Biofuels are gaining increased public and scientific attention, driven by factors such as oil price spikes, the need for increased energy security, concern over greenhouse gas emissions from fossil fuels, and government subsidies.

5.3.2.4 Energy Plantation

Large scale plantation of trees for production of energy from biomass is known as energy plantation. Latex of trees of the family *Euphorbiaceae e.g.* Jatropha and Euphorbia are used for energy plantations. Biodiesel is produced from the oil of Jatropha. Plantations on rural wastelands also help local population self-sufficient in their energy requirements.

5.3.3 Conventional (Non-Renewable) Energy Resources

5.3.3.1 Fossil Fuels or Fossil Fuel Based Energy

Fossil fuels are found inside the earth's crust where they have formed through heat and compression of forests, waste and other organic matter, which got buried due to earthquake, landslide *etc.*

Classification of Fossil Fuels

(A) Solid *(e.g. Coal)*

Coal is the most abundantly found fossil fuel in the world. It contains carbon, water, sulphur and nitrogen. Coal meets 70 per cent of the total energy needs of the

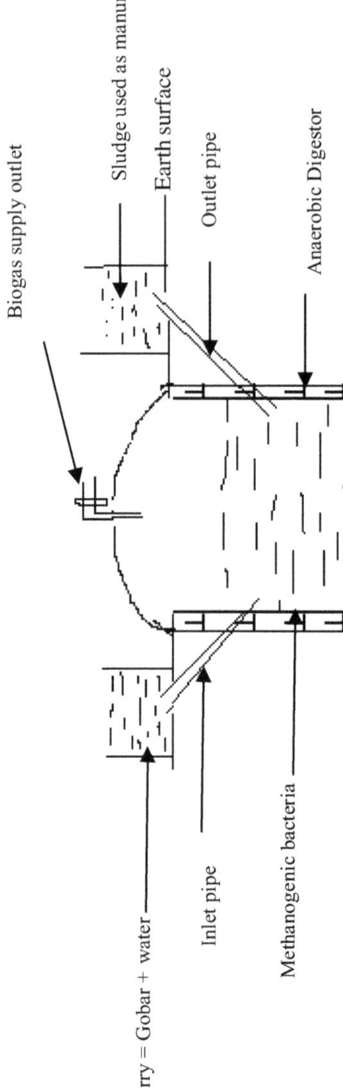

Figure 5.8: Anaerobic Decomposition of Biomass for Biogas Production (*MTU, 2012*).

Table 5.3: Generation, uses and environmental impacts of biofuels

Biofuel	*Generation*	*Uses*	*Environmental Impacts*
Bioalcohols (Bioethanol)	Action of microorganisms and enzymes through the fermentation of sugars (wheat, corn, sugarbeets, sugarcane, molasses) or starches (potato and fruit waste)or cellulose. Most common biofuel in world.	Bioethanol is used as fuel for vehicles (USA and Brazil).	A gasoline additive to increase octane and improve vehicle emissions.
Biodiesel	Biodiesel is produced from oils or fats of various crop plants known as **petro crops**. *e.g. Jatropha.* Biodiesel is made by trans-esterification of animal fats, vegetable oils, soy, rapeseed, jatropha, mahua, mustard, flax, sunflower, palmoil, hemp, field pennycress, pongamia, pinnata and algae is the most common biofuel in Europe.	Fuel for vehicles in its pure form.	Less air pollution from diesel-powered vehicles. Pure biodiesel (B100) is lowest emission diesel fuel.
Green diesel or Renewable diesel	Obtained from a variety of oils including canola, algae, jatropha and salicornia Green diesel is made by traditional fractional distillation process of oils."Green Diesel" is popularly used in Ireland.	Used as fuels along with diesel.	Less pollution
Syngas	Mixture of carbon monoxide, hydrogen and other hydrocarbons, produced by partial combustion of biomass.	May be burned directly in internal combustion engines, turbines or high-temperature fuel cells.	Less toxic emissions

world found and 87.4 per cent of all commercial energy. In India about 58 per cent of commercial energy is obtained from coal and 38 per cent from petroleum along with natural gas.

Coal is used for cooking, heating, in industries and thermal power plants. Petroleum is useful for transportation, agricultural equipments and some industries. Natural gas is used both in cooking and in industries.

Types of Coal or or Coal Energy

Peat, Bituminous (soft coal, 50 per cent carbon), Lignite (70 per cent carbon, brown coal), Anthracite (90 per cent carbon, hard coal).

Advantages of Coal

(i) Coal is present in large amount (most abundant). So it can be used as an energy source.

Disadvantages of Coal

(i) Burning of coal, produces, SO_2 *i.e.* cause air pollution.

(ii) Release of CO_2 and SO_2 gas in the atmosphere cause Green House Effect and Global Warming.

(iii) In thermal power plants, burning of coal also generates large amount of flyash. Flyash is a toxic waste, contains toxic heavy metals.

(iv) Workers in the coal mines suffer from following lung diseases: Black-lung disease, Asthma, Bronchitis, Lung cancer.

(B) Liquid (e.g. Petroleum)

Petroleum or Crude Oil

The gaseous fuels are basically derived from petroleum. It is a natural, underground fossil energy resource. It is formed due to decomposition of micro plankton deposited upon the sea beds, lakes and rivers for million of years. The decomposition takes place by the action of bacteria, under lack of oxygen and also by catalytic cracking. It is also called crude oil.

Advantages of Petroleum

1. Liquid fuel (Petroleum) is easy to transport.
2. Liquid fuel (Petroleum) is comparatively cleaner.
3. They have made possible the introduction and development of newer means of transport.

Disadvantages

1. After extraction it causes contamination in the water when the leakage takes place.
2. Its burning produces CO_2 and enhances the green house effect.
3. All combustion processes produce the pollutants like NO, SO_2 CO, NO_2, CO_2, Smog.
4. Petroleum contributes to acid rain and urban pollution.

(c) Gaseous (e.g. Natural Gas)

☆ Natural gas is a fossil fuel.

☆ It is eco-friendly fuel.

It is a mixture of hydrocarbon gases trapped under the earth's surface. It is mainly consisting of methane (CH_4), Propane (C_3H_8) and Butane (C_4H_{10}). After processing it is transported to supply filling stations. Natural gas can be used in two different forms.

1. LPG (Liquefied Petroleum Gas)

It is the mixture of Propane, Butane and Ethane.

2. CNG (Compressed Natural Gas)

It is mainly methane (CH_4).

Advantages

1. It is a clean fuel, requiring little processing.
2. It can be readily transported.
3. Smog formation is less in its use.
4. It is cheaper than petroleum.

Disadvantages

1. It requires both high pressure and low temperature for compression.
2. Thick walled tanks are required for storage as it is stored at a high pressure.
3. Methane is a green house gas.
4. Leakage, in any case, is a serious threat to the environment.

5.3.3.2 Nuclear Energy (*MTU, 2013-14*)

It presents nuclear power. It is a highly developed alternative for energy production in place of coal. Nuclear energy can be derived by two processes: Nuclear Fission, Nuclear Fusion.

Nuclear Fission (Chain-Reaction Mechanism)

The heavy nucleus on bombardment with neutron splits into lighter nuclei (Barium and Krypton) releases large amount of energy. One a.m.u. (Atomic Mass Unit) of uranium-235 yield energy equal to burning of 15 metric tons of coal.

Nuclear reactors are the devices need to liberate energy from nuclear fuels, under controlled conditions. The output of these reactors is in the form of a high temperature fluid. This can be used in the generation of electricity or as a direct source of heat for intensive industries.

Advantages of Nuclear Energy

(i) Large amount of energy is produced in nuclear fission reaction.

Disadvantages of Nuclear Energy

(i) High safety measures are required in the operations of Nuclear Reactors.

(ii) Highly equipped (knowledge + equipments) engineers are required for operating nuclear reactors.

(iii) **Disposal of Radioactive Waste**: Radioactive waste is released from Nuclear Power plants. It cannot be buried inside the earth and under the groundwater.

(iv) Due to high safety measures, nuclear reactors are located in isolated areas. To reduce the risks. Nuclear power plants are far distance away from the industries that require high energy in the areas of high population.

(v) **Nuclear disasters**: Due to **uncontrolled nuclear fission reaction** in the nuclear reactor, explosion occurs. *e.g.* **Chernobyl nuclear disaster** in Chernobyl, 1986 and **Fukushima nuclear disaster,** 2011.

- Due to radioactive pollution of Iodine-131, Cesium- 134, Cesium 137, thousands of people died and suffered from variety of diseases *viz.* loss of hair, thyroid; blood cancers, nausea, anaemia and ulcerating skin.
- Large social and economic loss occurred.
- **Genetic Diseases:** Sudden change in the genes is known as mutation, causes genetic diseases. Atom bomb explosion in **Hiroshima Nagasaki**, Japan also caused genetic diseases and large socio-economic loss.

Nuclear Fusion Two light nuclei of hydrogen fuse to form a helium nucleus with a release of enormous (extremely large amount of energy).

5.4 Steps of Government to Conserve Energy

1. Government departments involved in energy conservation are given below:
   (i) Central Pollution Control Board (CPCB).
   (ii) Ministry of Environment and Forest (MoEF).
   (iii) Department of Science and Technology (DST)
2. **Jawahar Lal Nehru Solar Energy Mission**

   Government has launched solar energy mission to
   (i) Promote the use of solar energy.
   (ii) Provide funds to the person or organization doing research to promote solar energy.
3. Promotion of CNG vehicles.
4. Giving subsidy on electricity operated Yo-Bikes in Delhi.
5. Afforestation programmes – Government is actively planting trees at road side known as avenue plantation.
6. Funding to research organizations involved in the research and increasing the applications of renewable energy resources *e.g.* TERI (Tata Energy Research Institute), NEERI (National Environmental Engineering Research Institute).
7. Mandatory (compulsory) energy audits (inspections) in the industries.

Participatory Learning

(a) Objective Questions

1. Biodiesel is primarily obtained from______.
2. The conventional energy resources are ______.
3. The non-conventional energy resources are ______.
4. Fuel cell in order to produce electricity burns ______ (*UPTU, 2008*).
5. Biogas is mainly ______.
6. The major purpose of river valley projects is ______.

True/False

7. Nuclear power plants are the source of thermal pollution.
8. India has a great potential of biomass based energy.

(b) Very Short Answer Type Questions

1. What is energy crisis?
2. Describe population explosion.
3. What are perpetual energy resources?
4. What is a solar cell or photovoltaic cell?
5. What is energy plantation?
6. What are the components of CNG, Natural Gas and LPG.
7. What are the advantages and disadvantages of conventional energy sources?
8. What are the advantages and disadvantages of non-conventional energy sources? (***MTU, Carryover, 2013***)

(c) Short Answer Type Questions

1. Write short-notes on
   (a) Nuclear power (b) Cold power.
2. Write short-notes on (***MTU, Carryover, 2012-13***)
   (a) Geothermal energy (b) Tidal energy
3. Describe the list of biofuels, their manufacturing and importance.
4. Write a short note on fuel cell. (***UPTU, 2009-10***)

(d) Long Answer Type Questions

1. What are the advantages and disadvantages of hydro energy (river-valley projects) and wind energy (wind-mill)?
2. Explain the ill-effects of mineral extraction. (***UPTU, 2010***)
3. Discuss the energy scenario with respect to Indian conditions. (***UPTU, 2006***) or how biogas can be generated.
4. What are the different types of biogas plants.
5. Differentiate among perpetual, conventional and non-conventional energy resources.
6. Describe potential of hydrogen as a future source of energy. (***UPTU, 2007***)
7. Write a short note on biofuels: syngas, biodiesel, biogasoline, bioethers and biogasohol.
8. What is meant by sustainable energy supply system? How it can be developed? What are the advantages and disadvantages?

Answers to Objective Questions

1. Jatropha
2. Non-renewable energy resources
3. Renewable energy resources
4. H_2
5. Methane
6. Electricity production
7. True
8. True

Chapter 6

Sustainable Development

6.1 Sustainable Development

Brudtland Definition of Sustainable Development

World Commission on Environment and Development (WCED) (The Brundtland Commission) in 1987, in its report to United Nations defined sustainable development as development carried out in such a way, the needs of present generation are fulfilled without compromising the ability of future generation to meet their own needs. Thus natural resources are conserved (saved) for future generation. Simply, it can be defined as development carried out in such a way, the needs of present generation are fulfilled and natural resources are conserved (saved) for future generation (**Figure 6.1**).

Development carried out in such a way that integrates social, economic and environmental factors is known as sustainable development.

Sustainable development = social benefit + economic benefit + environmental benefit

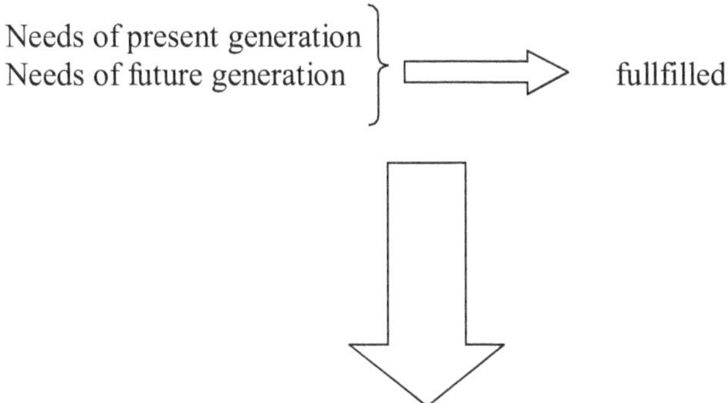

Figure 6.1: Objectives of sustainable development

6.2 Principles of sustainability

The principles of sustainability *i.e* necessary steps to achieve sustainable development (**Figure 6.2**) are as follows:

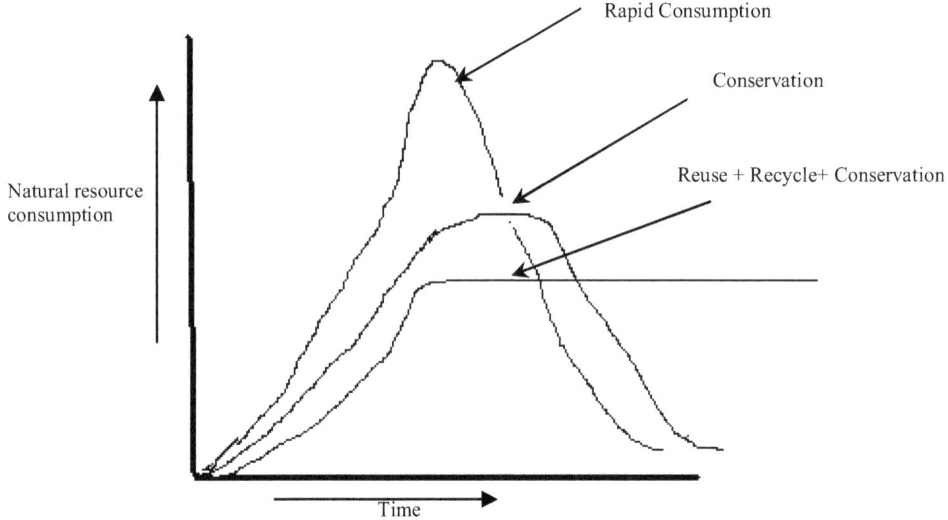

Figure 6.2: Principles of Sustainable Development.
(*Source*: Millers environmental sciences)

(i) Population Control

Population increases at exponential rate, puts pressure on natural resources, to fulfill the growing demands of population (air, water, soil, crops, forest, biodiversity, minerals) for fresh air, water, food, shelter *etc.* (**Figures 6.3a,b**).

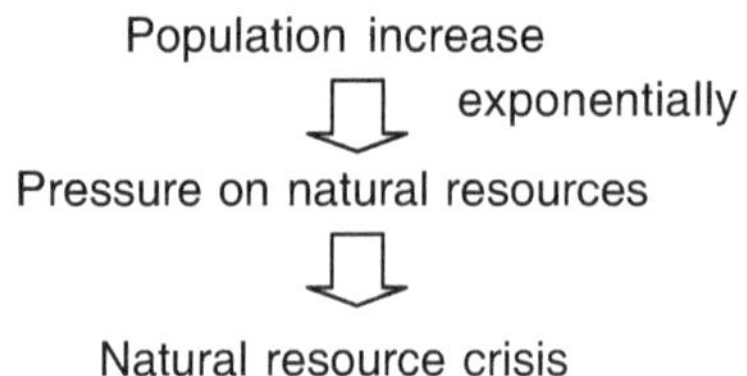

Figure 6.3a: Increase in Population Causes Natural Resource Crisis.

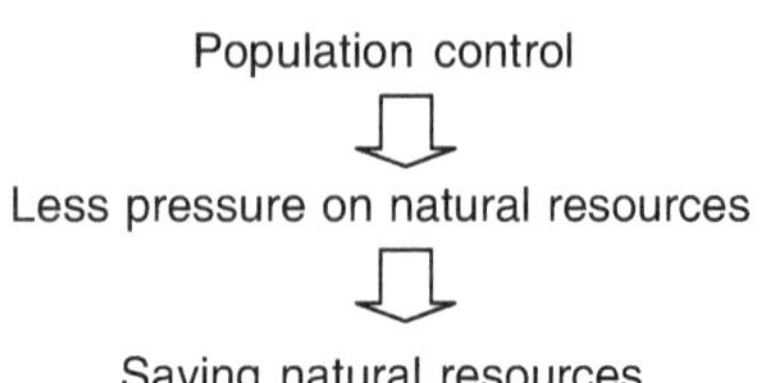

Figure 6.3b: Control of Population Saves Natural Resources

(ii) Conservation

Efficient use of natural resources, minimizing their wastage is the key to conservation.

(iii) Recycle

Recycling is an important step which reduces pressure on raw material from which the product is generated (closing the loop) (**Figure 6.4**).

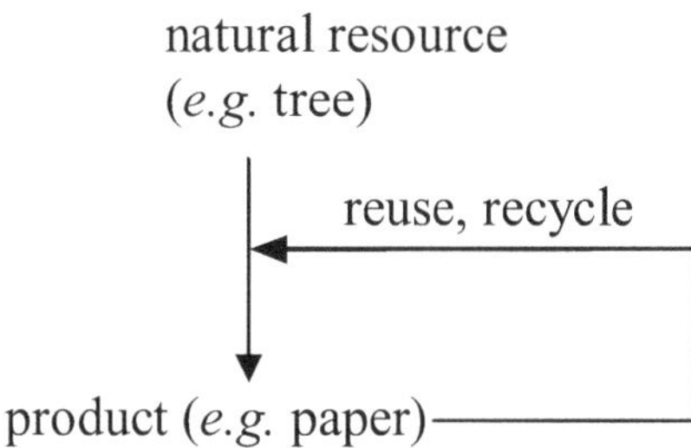

Figure 6.4: Recycling reduces the pressure on natural resources.

(iv) Wasteland Restoration or Wasteland Reclamation

Planting trees on wasteland *i.e.* revegetation is the most viable environmental option is known as Phytoremediation (Bioremediation by trees). The trees should be tolerant with respect to the pollution, fast growing and should contain high biomass (**Table 6.1**).

Table 6.1: Pollution Tolerant Trees.

Environmental Pollution	*Tolerant Trees*
Heavy metal (present in flyash, disposed in landfill dykes).	Arjun (*Terminalia arjuna*), *Bauhinia*, *Cassia*
Water Pollution	Water hyacinth removes heavy metals. *Pterris* sp. removes arsenic from polluted water.

6.3 Measurement of Sustainability or Sustainable Ethics or Equitable Utilisation of Natural Resources

(i) Intragenerational Equity

Development is carried out in such a way that needs of all the individuals of present generation are to be fulfilled.

(ii) Ecological Justice

All the individuals have a right to live in harmony with nature *i.e.* None of the living organisms should be disturbed.

(iii) Intergenerational Equity

Development is carried out in such a way that needs of present and future generation are fulfilled.

6.4 Sustainable Lifestyle (Role of an individual in sustainable development) (*MTU Carry, 2012-13*)

The key feature of sustainable lifestyle is saving natural resources in daily life. Life of Mahatma Gandhi is an example of sustainable lifestyle (*i.e.* conserving resources by minimizing wastage). *e.g.*

(i) Switching-off electricity when not needed.
(ii) Saving water in daily activities (water conservation).
(iii) Car-pooling
(iv) Reuse, Recycle
(v) Minimum use of polythene
(vi) Separating biodegradable and non-biodegradable waste at home.

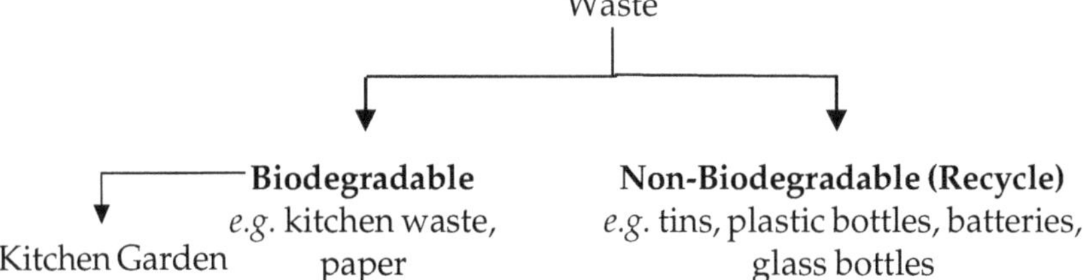

6.5 Challenges to Sustainable Development

1. Social Challenges

Consumerism

In today's scenario, the growing population has a range of lucrative products in the market, which they get attracted to buy known as consumerism. Media displays the products of different categories by advertisements. The consumers keeps buying the variety of products more than their actual necessity (**Figure 6.5**).

2. Challenges at Policy Level

Government has to develop an eco-friendly policy in terms of laws and legislations (**Figure 6.8**) with following major aspects:

(i) Increase in price of non-renewable energy resources *e.g.* petroleum, diesel.
(ii) Providing subsidy on cleaner fuels *e.g.* CNG, Natural gas.
(iii) Heavy price on car-parking.
(iv) Reduce the price of public transport.
(v) Heavy taxes on luxury items *e.g.* four wheeler, air conditioners, cosmetics, jewellary.
(vi) Real cost of water and electricity should be charged.
(vii) Polluting industries should pay heavy penalties.
(viii) Subsidy should be given on vehicles powered with renewable energy.

Market is flooded with many lucurative products, made up of non-biodegradable matter toys, stationery items, tiffins *etc.*

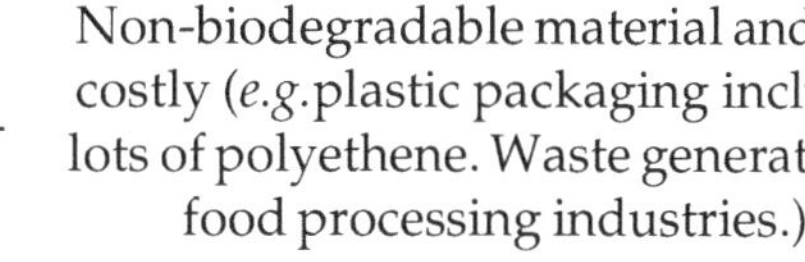

Non-biodegradable material and less costly (*e.g.* plastic packaging includes lots of polyethene. Waste generated in food processing industries.)

Consumerism: Consumer is compelled to buy lucrative products.

Global lifestyle has shifted towards consumerism or unsustainable consumption rather than sustainable lifestyle.

Pressure on natural resources

Rapid consumption of natural resources

Natural resource crisis *e.g.* ground water depletion, biodiversity loss, power crisis, crisis of fossil fuels (coal, petroleum, natural gas).

Here lies the challenge to make people aware about **environmental conservation** or to achieve **sustainability.**

Figure 6.5: Need to Aware the Consumer.

(ix) Delhi government is providing subsidy to electricity run YO-bikes.

Promoting residential buildings with rainwater harvesting systems, in new launching schemes.

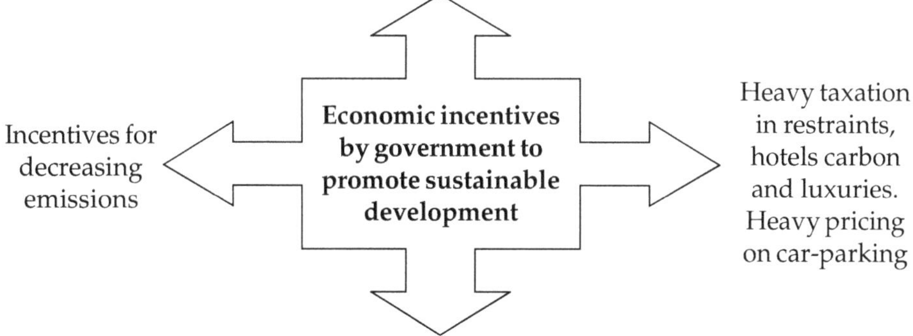

Figure 6.6: Some Illustrations Explaining Economic Incentives taken by Government to Achieve Sustainability.

3. Economic Challenges

The cost of the product should be decided (**Figure 54**) on the following factors:

(i) Economic Cost

The amount of expenditure occurred in terms of energy and monetary consumption is known as economic cost.

(ii) Manufacturing Cost

It involves the potential of the labour in manufacturing of a product.

(iii) Environmental Cost

The amount of natural raw material consumed and adverse environmental impacts of manufacturing and disposal of waste products cause depletion of natural resources. The extent of depletion of natural resources and compensation to reduce the impact of environmental pollution decides the environmental cost.

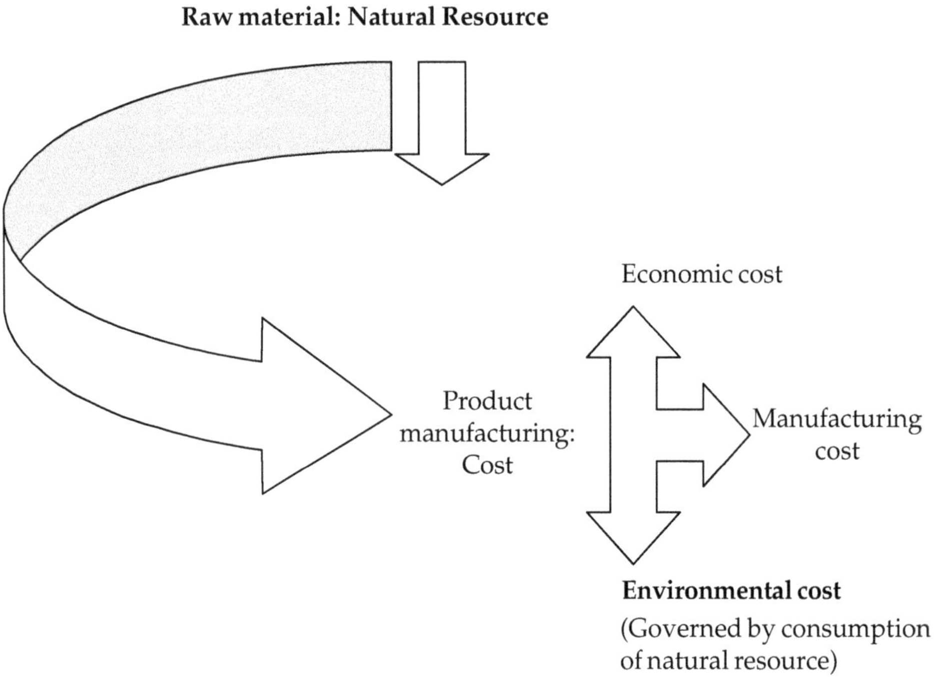

Figure 6.7: Factor Determining Cost of the Product.

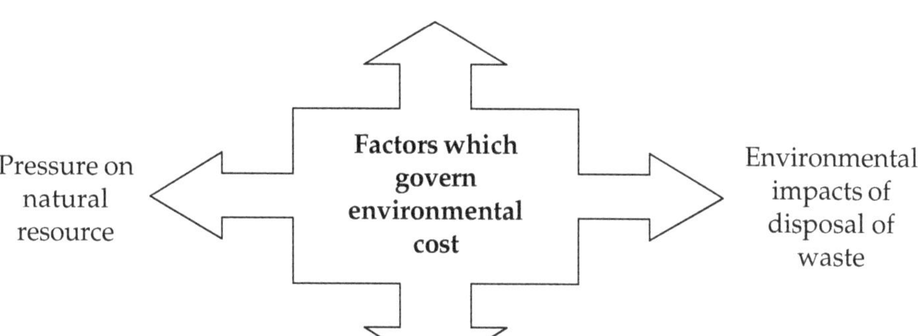

Figure 6.8: Factors Determining Environmental Cost.

6.6 International Efforts to Achieve Sustainability

(i) United Nations Conference in Environment and Development (UNCED) or Earth Summit

Agenda 21 is the list of principles or voluntarily accepted action plan of United Nations to achieve socio-economic and environmentally viable development in 21st century, in United Nations Conference in Environment and Development (UNCED), popularly known as earth summit held in Rio de Janeiro, Brazil in 1992. Major Themes of Agenda 21 include the quality of life on earth, efficient use of the natural resources, management of human settlements, the management of waste, sustainable economic growth

Important Areas of Environmental Conservation

Sustainable agriculture, water, energy, biodiversity, and biotechnology, forests, deserts and drought and mountain ecosystems are given special importance.

Important principles of Agenda 21 for reducing the green house gas emissions are as follows:

Social and Economic Dimensions

(a) Poverty removal, especially in developing countries.

(b) Promoting options for sustainable health.

(c) Promoting sustainable lifestyle.

Conservation and Management of Natural Resources

(a) Promoting population control.

(b) Control of air, water, land and pollution.

(c) Stopping deforestation.

(d) Conservation of biodiversity.
(e) Management of biotechnology and radioactive wastes.
(f) Protecting fragile environments.

Strengthening the Functions of Major Groups

(a) Strengthening the role of children and youth and women.
(b) Strengthening the role of non-governmental organizations.
(c) Strengthening the role of local authorities, businessmen and workers.
(d) Strengthening the role of indigenous (traditional) people, communities and farmers.

Tools for sustainable implementation of important steps

(a) Science, technology transfer, education and financial mechanisms.

(ii) Stockholm Conference

Stockholm Declaration on the Human Environment 1972 is a landmark in international relations. It placed the issue of protection of biosphere on the official agenda of policy and law of the member states. Environmental law instruments that link the environment and human rights. It states that "Man has the fundamental right to freedom, equality and adequate conditions of life, in an environment of quality that permits a life of dignity and well being" (Principle 1). Environmental Protection and development were conceptualised as two sides of a coin inseparable from each other. Hence Environmental protection was an essential element of social and economic development. Principle 6 states for the discharge of toxic substances that can cause serious or irreversible damage to ecosystems must be halted. Principle 15 states that planning must be applied to human settlements and urbanisation with a view of avoiding adverse effects on environment. Principle 18 incorporates the "precautionary principle" which propagates the avoidance of environmental risks.

(iii) Kyoto Protocol

An international agreement held in Kyoto, Japan in December, 1997 by different countries to take required measures to reduce their overall greenhouse emission to a level at least 5 per cent below the 1990 level by 2008-2012.

(iv) Montreal Protocol

An international agreement signed by twenty seven industrialized countries in 1987 to ensure protection of ozone layer present in stratosphere. More than 175 countries have signed the agreement by emphasizing on following major steps:

(a) **Sustainable Lifestyle**

Less use of ozone depleting substances.

(b) Finding out a substitute of ozone depleting substances.

(v) World Summit on Sustainable Development (WSSD)

Johannesburg Earth Summit held in Johannesburg, South Africa from 26 August to 4th September, 2002 described following important points at international level:

(a) Poverty removal.

(b) Think globally and act locally.

(vi) Rio +20 Earth Summit

The world once again came together at Rio De Janerio in June 2012, popularly known as Rio +20 Earth Summit, to discuss issues around sustainable development but summit did not result in any concrete outcome and failed to recreate the history of 1992. However concern for environment is always there which is evident from various treaties, protocols and conventions time to time.

Participatory Learning

(a) Objective Questions

1. Sustainability aims at
   (a) Conserving natural resources
   (b) All of these
   (c) Care for needs of future generation
   (d) Fulfilling the needs of present generation

2. The life of Mahatma Gandhi is an example of ______.

3. The concept of Sustainable Development was first discussed in the international conference______ known as ______.

4. Sustainable development integrates ______, ______ and ______ benefits. It is carried out to fulfill the demands of ______ and saving natural resources for ______.

(b) Very Short Answer Type Questions

1. What is the concept of sustainability?

2. Describe Agenda-21.

3. Define sustainable development. When does the idea conceived? Discuss its concept (***UPTU, 2008, 2009***).

(c) Short Answer Type Questions

1. We celebrate ozone day, earth day, environment day *etc.* What is the basic agenda?

2. Describe the role of environmental education in sustainable development.

3. What are the measures of sustainability?

(d) Long Answer Type Questions

1. What are the strategies of sustainable development?

2. Define Sustainable Development. What are the principles? (***MTU, 2012; MTU, 2013-14***).

3. Explain the equitable use of resources for sustainable lifestyle. (***UPTU, 2006, 2009***).

4. Explain the concept of sustainable development. What are the major obstacles in the path of sustainable development in India? (***UPTU, 2006, 2008***).

Answers to Objective Questions

1. (b)
2. Sustainable lifestyle
3. United nations conference on environment and development, earth summit
4. Social, economic, environmental, present generation or individuals of present generation, future generation or individuals of future generations

Chapter 7

Environmental Pollution

7.1 Environmental Pollution

Any undesirable change in the physical and chemical properties of air, water or land which adversely affects living systems is known as environmental pollution. The basic understanding of environmental pollution lies with the fact that the man-made physical and chemical processes in the industries and waste disposal deteriorates the air, water and soil quality known as air, water and soil pollution. Environmental pollution also occurs naturally *viz.* volcanic eruptions, forest fires, sand-dunes and cyclones.

7.1.1 Air Pollution

Any undesirable change in the physical and chemical properties of air which adversely affects living systems is known as air pollution.

7.1.1.1 Sources of Air Pollution

Man-Made Causes of Air Pollution

- ☆ **Stationary combustion sources or Industries:** Release toxic gases to the atmosphere. *e.g.* Thermal Power Plants, textile industries, fertilizer industries, chemical and pesticide industries.
- ☆ **Mobile combustion sources or Automobiles:** Automobiles are the largest source of air pollution. Due to incomplete combustion of fuel in the engines, toxic gases are released to atmosphere.

Natural Causes of Air Pollution

- ☆ Volcanic eruptions, forest fires, dust storms, cosmic dust, pollen grains, hydrogen sulphite and methane from anaerobic decomposition of organic matter also cause air pollution.

7.1.1.2 Effects of Air Pollution

On Plants

SO_2 degrades green chlorophyll to brown pigment. This reduces photosynthesis (Table 7.11a).

Chlorophyll (Green pigment) $\xrightarrow{\textbf{Degradation}}$ **Phaeophytin (Brown pigment)**

Table 7.1a: Effect of Air Pollutants on Plants.

Pollutant	*Effects on Plants*
SO_2	Degradation of green leaves (chlorosis), Necrosis (killing of issues)
O_3	Premature aging, suppressed growth, necrosis, Degradation of chlorophyll
NO_2	Suppressed plant growth, Degradation of chlorophyll
Fluorides	Necrosis at leaf tip.
Ethylene	Leaf abscission and epinasty (down ward curling of leaf)
PAN	Suppressed growth, silvering of lower leaf surface.

On Animals

Human suffer from respiratory problems, heart diseases and skin problems (**Table 7.1b**).

Table 7.1b: List of Air Pollutants and their Effects.

Source of Air Pollution	*Air Pollutant*	*Effect on Animals*
Coal mining	Coal dust	Black lung disease
Cigarette smoking	Tobacco	Heart disease
Nuclear reactors	Radioactive dust	Genetic disorders

On Buildings

Acid Rain causes corrosion of marble of Taj Mahal.

On Climate

Rise in temperature of earth due to increase in the level of Green House Gases (GHG). (CO_2, NO, SO, O_3, CH_4 *etc.*) known as global warming.

7.1.1.3 Classification of Air Pollutant

As per the method of entry into the atmosphere:

(i) Primary Air Pollutants

They are emitted directly into the atmosphere from the identifiable source of pollution. *e.g.* CH_4, CO_2, CO, SO_2, NO_2, NO *etc.* (**Table 7.2**).

(ii) Secondary Air Pollutants

Pollutants formed in atmosphere, as a result of chemical reactions (may be photochemical), between any combination of air pollutants and natural component of the atmosphere.

Table 7.2: Primary Air Pollutants and their Effects.

Primary Air Pollutant	*Source of Air Pollution*	*Environmental Effects*
SO_2	Industries (Coal mining, Cu mining)	Plants: adverse effect on photosynthesis. Animals: respiratory problems, eye disorders.
CO	Automobiles	Carboxyhaemoglobin (animals)
NO	Automobiles	Lung disorders
Lead	Automobiles	Brain disorders
Benzene	Automobiles	Blood cancers
Pesticides	Industries	Cause death when inhaled in excess.

Classification according to the physical state of pollutant

(i) Gaseous Pollutant

Most air pollutants exhibit gaseous properties *i.e.* they tend to obey Gas laws.

(ii) Solid Phase: Particulate Matter

(a) Aitken nuclei (very small particles).

(b) Fine particles (< 2.5 μm in diameter).

(c) Coarse particles (> 2.5 μm in diameter): These are generated mainly by mechanical processes and combustion. They undergo chemical reaction with primary air pollutants to form secondary pollutant. *e.g.* O_3, PAN (Per-oxy-Acetyl-Nitrate).

7.1.1.4 Control Measures of Air Pollution

- **Bioremediation:** Removal of toxic substance from air, water and soil with the help of living beings is known as bioremediation. Planting air pollution tolerant trees is the bioremediation option to control air pollution *e.g.* Arjun tree, Neem Tree, Ashok Tree. The trees to be chosen for bioremediation should have following characteristics:
  - (i) Trees should be fast growing.
  - (ii) They should have high biomass potential to accumulate large amount of toxicants.
  - (iii) They should be locally growing trees to be able to adapt to the climatic conditions.
- Government is promoting YO bikes which run by electricity.
- **Clean fuels:** Eco-friendly fuels *e.g.* CNG and LPG are promoted by government under the project "Clean Delhi".

- ☆ High traffic areas should be made for from schools, hospitals and residential areas.
- ☆ Using low sulphur coal in thermal power plants.

Air Pollution Control Devices

(i) Setting Chambers

It is a simply constructed air pollution device, made up of a chamber where the carrier gas velocity is reduced, so as to allow the particles to settle out of moving stream due to gravity.

(ii) Cyclone Separator

Due to rapid cyclonic movement of the gas, the particles are thrown towards the wall by centrifugal force.

(iii) Scrubber/Wet Collections

Scrubbers utilize water or any other specific liquid, to remove particles and gases by absorption or adsorption.

(iv) Filters

The fumes are passed through a filter to remove particulate matter.

(v) Catalytic Converters

Catalysts provide an extended surface area to adsorb the particles as studied in adsorption chemistry.

7.1.1.5 Air Pollution Disasters

The air pollution disasters were associated with temperature inversions lasting for several days, causing high concentrations of smoke and SO_2. Calm conditions of anticyclones, fogs and inversions hinder the dispersion of air-pollutants. Accumulation of air pollutants is enhanced in coastal environments.

(i) In 1930 the Meuse valley in Belgium was trapped by the inversion in which pollutants accumulated for 5 days resulting in the death of about 60 people.

(ii) Due to total absence of controls on vehicle exhaust emission, central Krachici was foggy. Existing power stations appears to be permanently foggy. Existing power stations emit highly poisonous gases. Local SO_2 concentration are three times higher than WHO's safe estimates.

(iii) Sudbury basin area, Canada is particularly rich in Ni-ore deposits. Thus world's important Ni-source, became a source of SO_2 pollution from Ni-smelting processes.

(iv) **Smog**: *Smog is a mixture of smoke and fog present in suspended droplet form.* There are two types of smog:

(a) London Smog (Grey smoke or Coal induced sulphurous smog)

It is combination of smoke (generated from coal combustion) and fog, responsible for poor atmospheric visibility, bronchial irritation and acid rain. The fog contains mainly a mixture of sulphur-dioxide, sulphur-trioxide and humidity (**Figure 7.1**).

(v) Similar stable meteorological conditions occurred in 1948 in Donora, USA.

London smog = smoke (SO_2, SO_3, Humidity) + fog

Coal burning (in urban areas, mainly at power plants)

Oxides of Sulphur + Particulate Matter

Certain meterological conditions: Oxidation (Sunrays), Stagnant air, stable air, thick fog, cloud cover, formation of inversion-layer

Concentrated sulphurous smog

Figure 7.1: Formation of London Smog.

Air Pollution Episode of London Smog

Date: December 5-9, 1952

Meterological conditions: Stagnant air, stable air, Thick fog, cloud cover, formation of inversion-layer (for continuous 5 days)

Effect of London Smog

(i) Death: 4,000 lives dead.

(ii) Peak SO_2 concentration was 1.3 ppm.

(iii) Smoke concentration was 4 mg/m^3 of air

(iv) Diseases: Lung diseases, Bronchitis, Pneumonia.

(v) Effect to plants: Leaf Necrosis- SO_2 cause death of the leaf tissue.

(b) Los Angeles Smog (Photochemical smog or Summer smog)

Oxides of carbon, nitrogen and sulphur, unburnt hydrocarbons (releases from automobile exhaust) and water vapors undergo a series of reactions in the presence of sunlight, to form a series of secondary pollutants *e.g.* unsaturated hydrocarbons, free radicals, organic peroxides and ozone. ***Peroxy-Acetyl Nitrate*** (PAN) is an important secondary pollutant and constituent of Los Angeles Smog.

$$NO_2(g) \xrightarrow{\text{UV radiations}} NO\,(g) + [O]$$

$$O_2(g) + [O] \xrightarrow{\text{UV radiations}} O_3$$

$$\text{Hydrocarbons} + \text{Oxides of Nitrogen} \longrightarrow O_3 + \text{PAN}$$

Peroxy-Acetyl Nitrate

In summers, when sunlight is available for longer hours, build-up of photochemical smog is easily possible.

Air Pollution Episode of Los Angeles Smog

Photochemical smog formation was first observed in Los Angeles in 1944.

Effect of Los Angeles Smog

(i) Hydrocarbons present in photochemical smog cause cancer.

(ii) Ozone and PAN cause eye irritation and watering.

(iii) It causes lung diseases and may lead to asthma attacks.

(iv) It may result into irritation to nose and throat.

(v) Damage to stomatal tissues may be caused.

(vi) Thousands of people died due to the episode of Los Angeles smog.

Control Measures of Smog

Smog is a result of air pollution. Thus, control of air pollution will help reduce smog and photochemical smog formation.

(c) Bhopal Gas Tragedy

Due to leakage of 40 tons of Methyl Iso Cyanate (MIC) from a union carbide factory during manufacturing of carbamate pesticide at the night of December 2, 1984 in Bhopal, thousand of people died due to its direct exposure and suffered from eye and respiratory disorders and lung cancers.

7.1.1.6 Long Range Transport of Gaseous Air Pollutants

The phenomenon are controlled by the density and strength of emissions of primary pollutants, prevailing meteorological conditions, transport and deposition processes and the rate of transformation from primary to secondary pollutants. Ozone is produced in the atmosphere and is highly reactive, thus its concentration can be reduced even as it is being produced. If the production rate is greater than consumption rate then high concentrations occur. This forms the important basis for the long range transport of measurable ozone concentrations.

The Meteorology of Long Range Transport

Importance of Long Range Transport (LRT) of air pollutants is in relation of the global air pollution problem.

(i) Regional scale transport occurs to distances of less than a thousand kilometers. Individual plumes merge and the distance allows a relatively uniform profile to develop after about one or two hours. It causes transport of pollutants from industrial and urban areas may affect the surrounding countryside downwind.

(ii) Sub-continental and continental scale transport occurs over hundreds to a few thousand kilometers.

(iii) Global air pollution transport moves from a few thousand kilometers to the entire atmosphere.

7.1.1.7 National Ambient Air Quality Standards (UPTU, 2010)

The standards for surrounding air content in the different regions is given in **Table 7.3**.

Table 7.3: National Ambient Air Quality Standards by CPCB (website: www.cpcb.nic.in).

Sl.No.	Pollutant	Time Weighted Average	Concentration in Ambient Air		
			Industrial, Residential, Rural and Other Area	Ecological Sensitive Area (Notified by Central Govt.)	Methods of Measurement
1	Sulful dioxide (SO_2), μg/m³	Annual* 24 hours**	50 80	20 80	☆ Improved West and Gaeke ☆ Ultraviolent fluorescence
2	Nitrogen dioxide (NO_2), μg/m³	Annual* 24 hours**	40 80	30 80	☆ Modified Jabo and Hochheiser (Na-Arsenite) ☆ Chemiluminescence
3	Particulate matter (size less than 10 μm) or PM30 μg/m³	Annual* 24 hours**	60 100	60 100	☆ Gravimetric ☆ TOEM ☆ Beta attenuation
4	Particulate matter (size less than 2.5 μm) or PM2.5 μg/m³	Annual* 24 hours**	40 60	40 60	☆ Gravimetric ☆ TOEM ☆ Beta attenuation
5	Ozone (O_3) g/m³	Annual* 24 hours**	100 180	100 180	☆ UV photometric ☆ Chemiluminescence ☆ Chemical method
6	Lead (Pb) μg/m³	Annual*	0.5	0.5	☆ AAS/ICP method after sampling on EPM 2000 or equivalent filter paper
		24 hours**	1	1	☆ ED-XRF using Teflon filter
7	Carbon Monoxide (CO) mg/m³	8 hours** 1 hour**	2 4	2 4	☆ Non Disperse Infra Red (NDIR) spectroscopy
8	Ammonia (NH_3) μg/m³)	Annual* 24 hours**	100 400	100 400	☆ Chemiluminescence ☆ Indophenol blue method

Contd...

Table 7.3–*Contd...*

Sl.No.	*Pollutant*	*Time Weighted Average*	*Concentration in Ambient Air*		
			Industrial, Residential, Rural and Other Area	*Ecological Sensitive Area (Notified by Central Govt.)*	*Methods of Measurement*
9	Benzene (C_6H_6) µg/m³	Annual*	5	5	☆ Gas chromatography based continues analyzer ☆ Adsorption and desorption followed by GC analysis
10	Benzo(a)Pyrene (BaP) - particulate phase only, mg/m³	Annual*	1	1	☆ Solvent extraction followed by HPLC/GC analysis
11	Arsenic (As), mg/m³	Annual*	6	6	☆ AAS/ICP method after sampling on EPM 2000 or equivalent filter paper
12	Nickel (Ni), mg/m³	Annual*	20	20	☆ AAS/ICP method after sampling on EPM 2000 or equivalent filter paper

*: Annual arithmetic mean of minimum 104 measurements in a year at a particular site taken twice a week 24 hourly at uniform intervals.

**: 24 hourly or 8 hourly or 1 hourly monitored values, as applicable, shall be compiled with 98 per cent of the time in a year. 2 per cent of the time, they may exceed the limits but not on two consecutive days of monitoring.

7.1.2 Water Pollution

Any undesirable change in the physical and chemical properties of water which adversely effects living systems is known as water pollution.

7.1.2.1 Sources of Water Pollution

- ☆ **Modern agriculture**: In modern agriculture, excess of chemical fertilizers and pesticides are supplied to the crops, which cause water pollution.
- ☆ **Industries**: The industrial waste is discharged into water – bodies. Toxic chemicals are present in industrial waste. *e.g.* Paper, Chemical, Food processing, Paint, Leather, Sugar, Industries, Mining.
- ☆ **Household activities**: Domestic activities also cause water – pollution.
- ☆ **Marine pollution**: Due to accidental leakage of oil during its transport, in the sea *i.e.* **oil spillage** and dumping of waste of the coastal cities in the sea are the major cause of marine pollution.

Categories of Sources of Water Pollution

(1) **Point sources:** Point sources are the identified single location of water pollution. *e.g.* sewerage systems, industrial effluent.

(2) **Diffused or non-point sources:** Their location cannot be easily identified *e.g.* the entry of scattered pollutants such as fertilizers and pesticides from a farm to nearby water-body.

7.1.2.2 Types of Water Pollutants (MTU, 2013-14)

Water pollutants are divided into three major categories: Physical, chemical and biological agents

(i) **Physical agents:** It mainly includes the physical factors *e.g.* heat released into water-bodies causes thermal pollution.

(ii) **Chemical agents:** Chemical wastes released into water bodies from chemical, metallurgical and food processing plants, textile paper and sugar mills, rubber and plastic industries, oil refineries, tanneries and other industries are categorized as below:

(a) **Organic chemicals**: *e.g.* Hydrocarbons and compounds containing benzene are hazardous and cause water pollution. Cellulose fibers, carbohydrates, proteins, oils, fasts, phenols, organic acids, aromatic compounds, antibiotic *etc.* are the examples.

(b) **Inorganic chemicals**: *e.g.* Arsenic, fluorides, nitrates are the source of arsenecosis, fluorosis and mithmaeglobaenemia respectively. Metals, chlorides, SO_4^-, CN^-, SCN^-, FeO_x, Cu, Cd, Hg and Cr acids and alkalis are the examples.

(c) **Fertilizers** *e.g.* Chemical fertilizers cause eutrophication in water-bodies.

(iii) **Biological agents:** Certain micro-organisms present in water are the source of water-borne and water-induced diseases. *e.g.* protozoa and bacteria may degrade water quality. *Escherichia coli* is the bacteria present in fecal matter, when enters in water body makes the water unfit for drinking.

7.1.2.3 Effects of Water Pollution

Various water pollutants are so toxic that they cause death of living organisms (due to their toxicity). Thus they come under the category of hazardous waste. The list of adverse effects of water pollution is given below:

- ☆ **Eutrophication**: In modern agriculture, chemical fertilizers are supplied to the crops. The water runoff with chemical fertilizers will reach to the nearby water-body. Small water plants grow in excess known as **algal bloom**. This prevents inter-mixing of atmospheric oxygen to dissolved oxygen in water. The water plants and animals start dying due to lack of dissolved oxygen. The dead parts will deposit at the bottom of the water body. The process is known as eutrophication. The water body is known as eutrophic water body. *e.g.* Chilka lake in Orissa (**Figure 2.2**).
- ☆ **Thermal pollution**: Due to discharge of industrial waste, temperature of water body increases.
- ☆ **Biomagnification:** The concentration of toxic chemical increase several times when it is transferred from one tropic level to another *e.g.* (**minamata disease** in Japan) the people died when consumed Mercury contaminated fish in Japan. Mercury containing chemical is accidentally released in the bay of Minamata in Japan immediate death of people occur known as minamata disease.
- ☆ The bad smell comes out of the polluted water body.
- ☆ Penetration of sunlight decreases in polluted water body, reducing the plant's ability to photosynthesize.
- ☆ **Biological Oxygen Demand (BOD):** The amount of dissolved oxygen required by micro-organisms to decompose the organic matter present in the waste water is known as Biological Oxygen Demand. BOD increases due to water pollution.

7.1.2.4 Water Quality Standards

Water quality criteria is explained by CPCB as showing in Table 7.4.

7.1.2.5 Control of Water Pollution

Wastewater Treatment

The waste water from industries should be treated before discharging into the water-bodies.

Bioremediation

Water hyacinth absorbs toxic chemicals from water-body, the process of detoxification with the help of living organisms is known as bioremediation.

Sustainable Agriculture

Integration of traditional and modern agricultural techniques reduces water pollution *e.g.* agroforestry, terrace-farming.

Table 7.4: Water Quality Criteria is Explained by CPCB.

Designated-Best-Use	*Class of Water*	*Criteria*
Drinking Water Source without conventional treatment but after disinfection	**A**	☆ Total Coliforms Organism MPN/100ml shall be 50 or less ☆ pH between 6.5 and 8.5 ☆ Dissolved Oxygen 6mg/l or more ☆ Biochemical Oxygen Demand 5 days 20°C 2mg/l or less
Outdoor bathing (Organised)	**B**	☆ Total Coliforms Organism MPN/100ml shall be 500 or less pH between 6.5 and 8.5 Dissolved Oxygen 5mg/l or more ☆ Biochemical Oxygen Demand 5 days 20°C 3mg/l or less
Drinking water source after conventional treatment and disinfection	**C**	☆ Total Coliforms Organism MPN/100ml shall be 5000 or less pH between 6 to 9 Dissolved Oxygen 4mg/l or more ☆ Biochemical Oxygen Demand 5 days 20°C 3mg/l or less
Propagation of Wild life and Fisheries	**D**	☆ pH between 6.5 to 8.5 Dissolved Oxygen 4mg/l or more ☆ Free Ammonia (as N) 1.2 mg/l or less
Irrigation, Industrial Cooling, Controlled Waste disposal	**E**	☆ pH between 6.0 to 8.5 ☆ Electrical Conductivity at 25°C micro mhos/cm Max.2250 ☆ Sodium absorption Ratio Max. 26 ☆ Boron Max. 2mg/l
	Below E	☆ Not Meeting A, B, C, D and E Criteria

7.1.3 Thermal Pollution

Discharge of hot industrial waste water (mainly from thermal power plants and nuclear power plants) into nearby water-body causes rise in the temperature of rivers or lakes that is injurious to aquatic life (**Figure 7.2**).

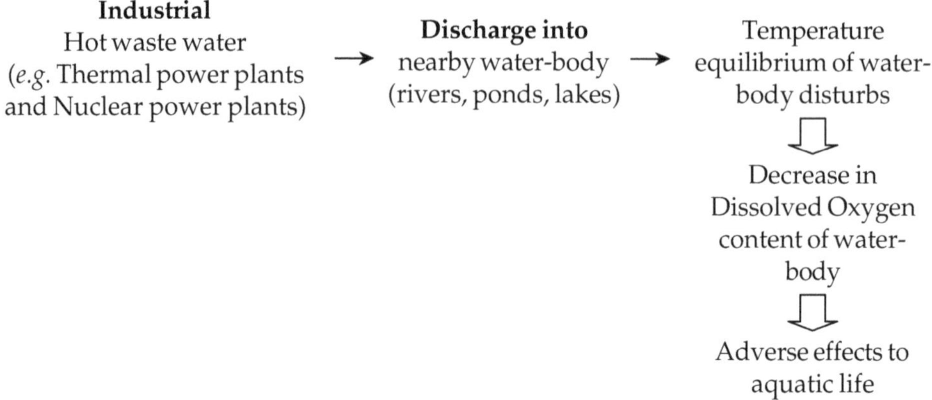

Figure 7.2: Causes and Effects of Thermal Pollution.

7.1.3.1 Sources of Thermal Pollution

Major sources that lead to thermal pollution can be classified into the following categories:

(i) Thermal Power Plants

Coal is burnt to generate electricity in thermal power plants.

(ii) Nuclear Power Plants

Water is used as a cooling agent in nuclear power plants.

7.1.3.2 Effects of Thermal Pollution

Health Effects

(i) Increase in water temperatures lower dissolved oxygen content by decreasing the solubility of oxygen in water.

(ii) Warmer water also increases respiration rates of aquatic organisms (fish, amphibians and copepods) and as a result they consume oxygen faster, and it increases their susceptibility to disease, parasites, and toxic chemicals.

(iii) At high temperature, photosynthesis is adversely affected.

(iv) Some aquatic animals migrate to suitable environments.

(v) Eggs of animals and insect larvae die at high temperatures of water.

Ecological Effects

(i) Primary producers are affected by warm water because higher water temperature increases plant growth rates, resulting in a shorter lifespan and species overpopulation. This can cause an algal bloom which reduces oxygen levels.

(ii) Changes in temperature also result in a migration of organisms to more suitable environment. This results in decrease in biodiversity.

7.1.3.3 Control of Thermal Pollution

Hot waste water should be cooled before discharging into the water-body by cooling ponds and cooling towers.

(i) Cooling ponds are the man-made water-bodies designed for cooling by evaporation, convection, and radiation.

(ii) Cooling towers are the instruments which transfer waste heat to the atmosphere through evaporation and or heat transfer.

(iii) Cogeneration is another process of recycling of waste heat for domestic and/or industrial heating purposes.

7.1.4 Waste Water Treatment Plant

Waste Water

Rejected water (of no further use) released from homes (domestic source) is known as sewage. Rejected water from industries is released into nearby water-bodies is known as effluent.

Removal of excessively accumulated waste matter from the waste water is carried out by waste water treatment plant. Lay out of waste water treatment plant is shown in **Figure 7.3**.

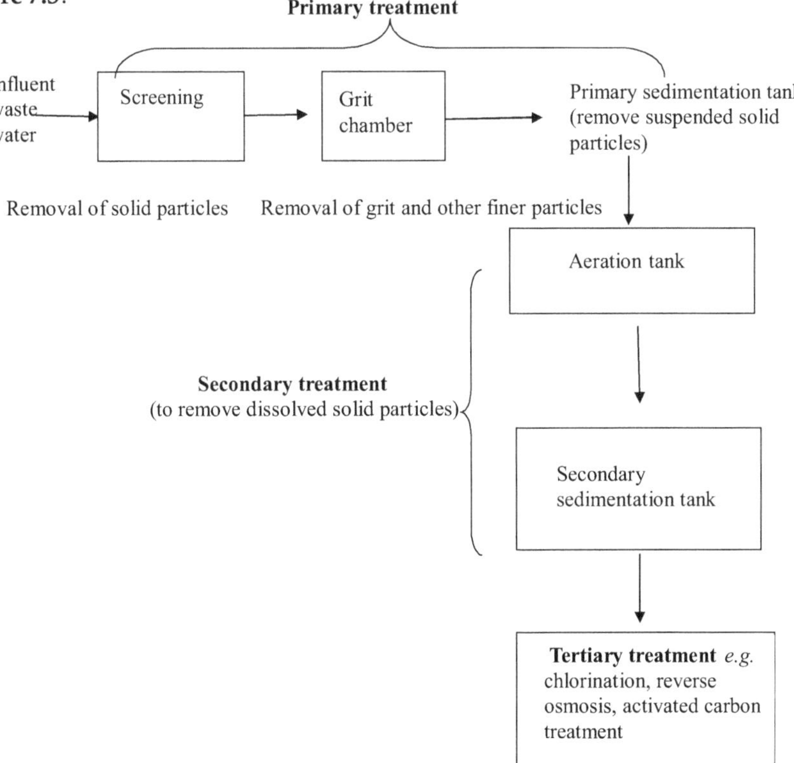

Figure 7.3: Layout Diagram of a Simple Waste Water Treatment Plant.

Steps of Waste Water Treatment Plant

(i) Primary Treatment (Physical Treatment)

It includes steps:

- ☆ **Screening:** Large suspended or floating particles are removed by screening.
- ☆ **Grit chamber:** Different sized particles are removed by grits of different sieve size.
- ☆ **Sedimentation:** Suspended solid particles settle down in the settling chamber.
- ☆ Thus, the amount of suspended solids is reduced. BOD is also reduced this way.

(ii) Secondary Treatment (Biological Treatment)

Decomposers breakdown complex organic matter (present in waste water) to simple form and help reducing BOD and suspended solids.

Types

1. Aerobic Treatment

Decomposers breakdown complex organic matter (present in waste water) to simple form, in the presence of oxygen under aerobic treatment.

$$\text{Complex organic matter (waste water)} \xrightarrow{\text{Decomposition}} \text{simple form}$$

(a) Trickling Filter

Bed of stones is covered with Biological film or Biological slime (Bacteria *e.g. Pseudomonas*, flavobacteria), fungus, algae, worms, insect larvae). Organic matter is decomposed by the microorganisms. Algae are the source of oxygen. It carries out photosynthesis in the presence of CO_2, H_2O and sunlight to make its own food and oxygen (**Figure 7.4**).

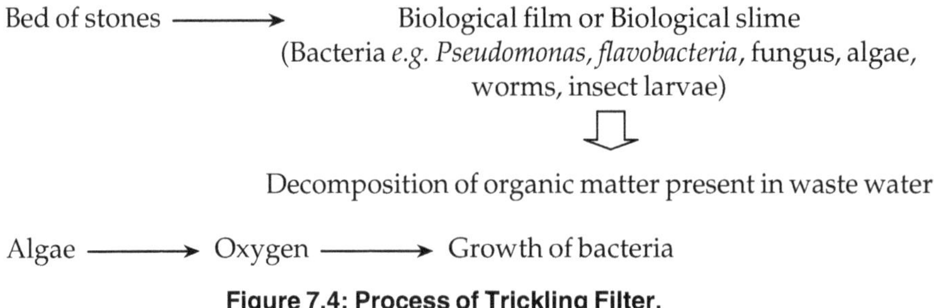

Figure 7.4: Process of Trickling Filter.

(b) Activated Sludge Process

Steps

(i) **Wastewater + oxygen:** Decomposers carried out oxidation of organic waste into simple form when oxygen is present in waste water.

(ii) Solid form of waste and group of bacteria is known as activated sludge.

(iii) **Solid-liquid separation:** Waste is separated in the solid and liquid forms. Solid waste is activated sludge and liquid waste is sent for tertiary treatment.

(iv) **Sludge re-cycle system:** Activated sludge is recycled to aeration tank for further aerobic decomposition.

(c) Oxidation-pond

The few meters deep pond where two zones are formed depending upon the penetration of sunlight. Aerobic zone is formed till sunlight penetration occurs. Anaerobic zone starts when sunlight penetration stops. Photosynthesis occurs in aerobic zone and bacterial decomposition occurs in anaerobic zone.

(d) Tertiary (Advanced or Chemical) Treatment

Removal of impurities from waste water (released after secondary treatment) by advanced chemical processes is known as tertiary treatment.

Reverse Osmosis (R.O.)

Movement of water from high solute concentration to low solute concentration, under pressure is known as reverse osmosis.

Nitrogen Removal

Decomposition: Decomposers breakdown ammonia present in waste water.

(i) **Nitrification**: Nitrifying bacteria convert ammonia to nitrite (*Nitrosomonas*) to nitrate (*Nitrobactor*).

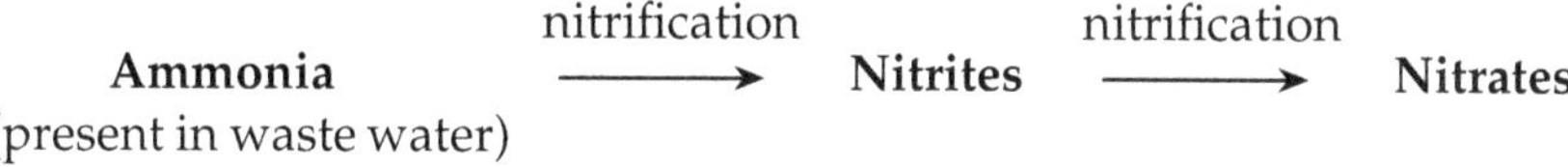

(ii) **Denitrification:** Denitrifying bacteria returns nitrogen (of nitrates) from waste water to atmosphere (in the absence of oxygen). *e.g. Pseudomonas.*

$$\underset{\text{(present in waste water)}}{\textbf{Nitrates}} \xrightarrow{\text{denitrification}} \underset{\text{(removal of nitrogen from wastewater)}}{N_2}$$

7.1.5 Soil Pollution or Land Degradation

Any undesirable change in the physical and chemical properties of land which adversely affects living systems is known as soil or land pollution.

7.1.5.1 Sources of Soil Pollution

- **Modern agriculture:** Use of chemical fertilizers and pesticides degrades soil fertility.
- **Industrial waste:** Disposal of solid or semi-solid industrial waste on land cause land degradation.
- **Radioactive pollution:** Radioactive substances are released from nuclear plants into soil emitting radiation.
- **Acid rain:** Acid rain increase the normal pH of soil.

7.1.5.2 Effects of Soil Pollution

- Soil pollution reduces soil fertility.
- It causes contamination of ground water.
- It adversely affects human health.

7.1.5.3 Control Measures

- **Sustainable Agriculture:** Integration of traditional and modem agricultural techniques reduce soil erosion. *e.g.*
  - (*a*) Use of **biopesticides** (neem).
  - (*b*) Use of **biotertilizers** (*e.g.* cow dung, alfala).
  - (*c*) **Agroforestry** (trees+crops) reduces the soil erosion because roots of trees keep the soil intact.
  - (*d*) **Terrace Farming**: Farming on steps reduces soil erosion in mountains. (discussed in detail in the section).
- **Bioremediation/Wasteland Reclamation/Wasteland Restoration**

  Planting tolerant trees remove pollutants from the soil *e.g.* Arjun, Neem, Bargad and Pipal tree are tolerant to heavy metals.
- **Solid Waste Management**
  - (*a*) Recycling of solid waste.
  - (*b*) Reuse, recycle and recovery are important methods.
  - (*c*) Biogas production from biodegradable waste.
  - (*d*) Safe disposal of waste (landfill).

7.1.6 Solid Waste

Definition

Rejected material (which is of no further use) in solid form from houses. Commercial and industrial sector, mining and agricultural activities cause environmental pollution (air, water, land and ground water pollution).

7.1.6.1 Sources of Solid Waste

(i) **Domestic:** Waste coming out how house-holds is known as municipal solid waste. It includes biodegradable and non-biodegradable waste.

(ii) **Industries:** Waste generated from various industries is known as industrial waste (**Table 7.5**).

(iii) **Hospitals:** *e.g.* bones, used cotton and syringes

(iv) **Offices:** *e.g.* paper.

Table 7.5: Industrial waste and environmental effects

Industry	*Solid Waste*	*Adverse Environmental Effects*
Packaging industry	Plastic	Plastic is non-biodegradable, reduces soil fertility, develops phobia in roots of plants and may cause intestinal cancers in animals if ingested.
Thermal Power Plants	Flyash	Flyash contains toxic heavy metals, generated in tons. It causes air, water and land pollution when spread away from disposal sites. Leaching of heavy metals to groundwater, causes ground water pollution.
Cement industry	Silica dust	It also causes air, water, land and ground water pollution. Workers may suffer from silicosis, a lung disease; since cause is workplace, it is known as occupational disease.
Mining	Metal waste	Mining causes air, water, noise, land and groundwater pollution.

7.1.6.3 Effect of Solid Waste

Municipal solid waste is a major environmental problem in developing countries, especially in India, discussed below:

(i) Separating biodegradable and non-biodegradable wastes in separate dust-bins at the time of their generation is known as segregation at source. This practice is lacking at most of households *i.e.* Most of the individuals do not throw biodegradable and non-biodegradable waste in separate dust-bins in their houses.

(ii) Plastics thrown here and there with wasted food materials are eaten up by cows, which may stick to their intestine and cause cancers.

(iii) Uncollected MSW gets its way in the drains, which may be blocked, resulting in flooding and insanitary conditions (water-borne and water induced diseases). *e.g.* Plague spread in Surat, India 1994.

(iv) Hazardous chemicals *e.g.* Furans and dioxins are released due to open burning of wastes.

(v) Aerosols and dusts can spread fungi and pathogens from uncollected and decomposing wastes.

(vi) Leachate flowing from waste dumps and disposal sites cause ground water pollution.

(vii) Chemical wastes (especially persistent organics) may be fatal.

(viii) Fires on disposal sites can cause major air pollution, causing illness and reducing visibility and making disposal sites dangerously unstable.

(ix) Open-dumping of solid wastes is most common in most of the cities of India. Toxic gases entering in the atmosphere may cause air-pollution. Workers in the dumpsite may suffer from lung and skin diseases. Toxic gases may transform in the atmosphere and precipitate down on the land and water surfaces, causing land and water pollution. Leaching of toxicants from the dumpsite may cause groundwater-pollution.

7.1.6.3 Control Measures of Solid Waste (MTU, 2013-14)

Solid Waste Management

(1) Five Rs (Refuse, Reduce, Reuse, Recycle, Recovery)

(i) **Refuse:** Purchasing only the required amount of items, inspite of lucrative offers available in market reduces the unnecessary consumption or utilization and waste generation.

(i) **Reduce:** Efficient use of natural resources reduces waste generation. (*i.e.* less wastage).

(ii) **Reuse:** Using the waste material for another purpose without any processing (**Table 7.6).**

Table 7.6: Reuse of waste materials

Waste Material	*Re-Use*
Plastic Bottles (Cold drink *etc.*)	Water-Bottle
Tins	Pencil stands
One Side vacant copy	Rough note-book

(iii) **Recycle:** The non-biodegradable waste is sent to processing unit to form the product. *i.e.* waste is considered as raw material so it reduce pressures on natural resource.

(iv) **Recovery:** The useful material is extracted from the waste *e.g.* metal.

(2) Disposal of Solid Waste

(i) A landfill design is suggested to dispose solid waste safely. The ultimate repository of a city's MSWM is well designed landfills which are in accordance with appropriate local health and environmental standards. Waste is deposited in 0.9-4.5m thick layers in depressions and then compacted and covered at least once a day by earth with bulldozers. Landfills are expensive to operate as there are requirements concerning daily cover, liners, leachate collection, gas collection, monitoring, hazardous waste exclusion, closure and post closure requirements and financial assurances. Selecting the site, certain restrictions *e.g.* water supplies, endangered or threatened species, scenic rivers, recreation or preservation areas utility or transmission lines. Sanitary landfill design is a complex process involving disciplines such as geomechanics, hydrology, hydraulics, wastewater treatment and microbiology.

(ii) **Composting:** Biodegradable waste can be biodegraded to manure with the help of bacterial decomposition forming compost. Anaerobic composting is a slow process, for 4-2 months, carried out at low temperature in the absence of oxygen and produces offensive odor. Aerobic compost is formed rapidly in the presence of oxygen at high temperature with bad odors. MSW offers good possibilities for recovery of energy in its organic fraction for gainful utilization and reduction of total quantity of waste by nearly 60 per cent to over 90 per cent cost of transportation and demand for land and net environmental pollution also get reduced.

(iii) **Vermicomposting:** Earthworms also degrade biodegradable waste forming nutrient rich manure known as vermicompost.

(iv) **Anaerobic Digestion or Biomethanation or Biogas formation:** The latent energy present in the organic fraction of MSW is recovered by bioconversion by methanogenic bacteria under anaerobic conditions and very high temperature (preferable high percentage of organic biodegradable matter and high moisture content) discussed in section.

(v) **Incineration/pyrolysis (Thermo-chemical conversion or gasification)** (preferably low moisture containing wastes): Non-biodegradable toxic waste is treated at high temperature and pressure reducing its 90 per cent volume and 75 per cent weight. The primary products of combustion are carbon-di-oxide, water vapour, nitrogen and solid residue of glass, ceramics, mineral ash *etc.* Since it is the costly process, incineration method is adopted for disposing mainly hazardous waste.

(vi) **Disposal of Solid Waste:** Solid waste disposal involves following important points:

1. The disposal site should be selected far away from ecologically important areas *e.g.* water supplies, endangered or threatened species, scenic rivers, recreation or preservation areas utility or transmission lines.
2. A landfill design is suggested to dispose solid waste safely with appropriate local health and environmental standards. Developing a sanitary landfill design involves disciplines such as geomechanics, hydrology, hydraulics, wastewater treatment and microbiology. Waste is disposed in 0.9-4.5m thick layers in depressions and then compacted and covered at least once a day by earth with bulldozers.
3. Operations followed in a landfill are expensive *viz.* daily cover, liners, leachate collection, gas collection, monitoring, and hazardous waste exclusion.

A reliable data generation system of MSW of waste generation, collection and storage, transportation and treatment and disposal facilities of Indian cities is an important step towards sustainability. There is a strong need for sustainable municipal solid waste management encompassing planning, engineering, organization, administration, financial and legal aspects of its basic component in environmentally compatible manner *i.e.* adopting the principles of economy, aesthetics, energy and conservation.

7.1.7 Status of Solid Waste Management in India

Urban solid waste management in India is the most neglected area (**Table 7.7**), discussed as follows:

(i) No system of segregation of organic, inorganic and recyclable wastes exists at the household level.

(ii) In most of the Indian cities and towns, waste from hospitals and nursing homes is also coming in MSW. However, as per legislation it is required to be collected and treated separately.

(iii) In India, wastes are normally high in biodegradable matter and low in paper, metal and glass, vary seasonally and place to place depending on lifestyle, food habits, standard of living and degree of commercial and industrial activity.

(iv) MSW is dumped on the low-lying dumping sites on the outskirts of the city. Open dumping of garbage caused serious health problems, ground water contamination by leachate production and emission of atmospheric pollutants.

7.1.8 Hazardous Waste Management (HWM)

A solid waste or combination of solid wastes that, because of quantity, concentration or physical, chemical or infectious characteristics, may cause or significantly contribute to an increase in mortality or an increase in serious, irreversible, or incapacitating reversible illness or pose a substantial hazard to human health or the environment when improperly treated, stored, transported, disposed or otherwise managed *i.e.* Lethal cause immediate death is known as hazardous waste.

1. Categories of Hazardous Waste and their Characteristics in HW (Management and Handling) Rules, 1989

In order to manage hazardous waste (HW), mainly solids, semi-solid and other Industrial wastes, Government of India notified the Hazardous Waste (Management and Handling) Rules (HWM Rules) on July 28, 1989 under the provisions of the Environment (Protection) Act, 1986 (Section 6, 8 and 25) and amended in the year 2000 and 2003 (**Table 7.8**). These amendments are to identify hazardous wastes in Schedule I and also by way of concentrations of specified constituents of the hazardous waste in Schedule II. Categories of wastes banned for export and import are given in Schedule-8. The procedure for registration of the recyclers/reprocessors with environmentally sound facilities for processing waste categories such as used lead acid batteries, non-ferrous metal and used oil as contained in schedule-4 and schedule-5 respectively are given. Further, separate Rules have also been notified in continuation of the above Rules for bio-medical wastes as well as used lead acid batteries.

2. Storage

Large quantities of hazardous waste are stored in on-site tanks.

Table 7.7: Solid Waste Management Scenario in India.

Composition	*Collection and Transport*	*Treatment*	*Disposal*	*Legislative Framework*	*Proposed Solid Waste Management India (Recommendations)*
Biodegradable content: higher; ash and fine earth: high.	Poor collection efficiency (i) less manpower, containers and transportation facilities (ii) uncovered transportation.	(i) Manual separation of waste by the generators and ragpickers. (ii) Recycling of non-biodegradable waste. (iii) Composting of bio-degradable waste. (iv) A biomethanation plant was proposed at Pune and Mumbai, but its viability is yet to be proven. A project for producing 105 tpd fuel pellets for municipal solid wastes (MSW) in Hyderabad has been installed. Work on a four megawatt MSW-based power plant in Nagpur has commenced.	Landfill disposal is not popular: increasing population, rising cost. Dumping: low-lying areas on the outskirts of the city causes environmental pollution.	*National policy and l egislation for MSWM, Municipal Solid waste (Management and Handling) Rules, 2000*: (i) prohibiting littering of street, (ii) house to house waste collection (iii) awareness programmes (iv) adequate community storage facilities, (v) use of colour code bins and (vi) waste segregation, (vii) covered transport (viii) composting, anaerobic digestion, pellatisation and upgrading of existing dump-sites and disposal of inert wastes in sanitary landfills.	(i) Vermicomposting (ii) *Research organisations*: TEAM, New Delhi (TERI Enhanced Acidification and Methanation, to generate high quality methane-rich fuel); NEERI (National Environmental Engineering Research Institute, Nagpur). (iii) **GIS based mapping of area**: NDMC (New Delhi Municipal Corporation) has installed computer-aided design software and high definition scanning devices, to improve SWM.

Table 7.8: 18 Categories of Hazardous Waste.

Waste Categories	*Type of Waste*	*Regulatory Quantities*
Waste Category No. 1	Cyanide waste	1 kilogrammes per year calculated as cynide
Waste Category No. 2	Metal finishing waste	10 kilogrammes per year the sum of the specified substance calculated as per metal
Waste Category No. 3	Waste containing water soluble chemical compounds of lead, copper, zinc, chromium and antimony	10 kilogrammes per year the sum of the specified substance calculated as per metal
Waste Category No. 4	Mercury, Arsenic, Thallium and Cadmium bearing wastes.	5 kilogrammes per year the sum of the specified substance calculated as pure metal.
Waste Category No. 5	Non-halogenated hydrocarbons including solvent.	200 kilogrammes per year calculated as non-halogenated hydrocarbons.
Waste Category No. 6	Halogenated hydro-carbon including solvents	50 kilograms per year calculated as helogenated hydrocarbons.
Waste Category No. 7	Wastes from paints, pigments, glue, varnish and printing ink.	250 kilogrammes per year calculated as oil or oil emulsions.
Waste Category No. 8	Wastes from Dyes and Dye intermediate containing inorganic chemical compounds.	200 kilogrammes per year calculated as inorganic chemicals.
Waste Category No. 9	Wastes from Dyes and Dye intermediate containing organic chemical compounds.	50 kilogrammes per year calculated as organic chemicals.
Waste Category No. 10	Waste oil and oil emulsions.	1000 kilogrammes per year calculated as oil and oil emulsions.
Waste Category No. 11	Tarry wastes from refining and tar residues from distillation or prolytic treatment.	200 kilogrammes per year calculated as tar
Waste Category No. 12	Sludgesarising from treatment of waste waters containing heavy metals, toxic organics, oils emulsions and spend chemical and inceneration ash.	Irrespective of any quantity.
Waste Category No. 13	Phenols.	5 kilogrammes per year calculated as phenols.
Waste Category No. 14	Asbestos.	200 kilogrammes per year calculated asbestos.
Waste Category No. 15	Wastes from manufacturing of pesticides and herbicides and residues from pesticides and, herbicides formulation units.	5 kilogrammes per year calculated as pesticides and their intermediate products.
Waste Category No. 16	Acid/Alkaline/Slurry	200 kilogrammes per year calculated as Acids/Alkalies.
Wastes Category No. 17	Off-specification and discarded products.	Irrespective of any quantity.
Wastes Category No. 18	Discarded containers and Containers linears of hazardous and toxic wastes.	Irrespective of any quantity

3. Transport

Safe transport is ensured to disposal sites under covered tankers, trucks or rail cars.

4. Spillage

An emergency plan for accidental release of hazardous pollutants.

5. Disposal

Hazardous waste is disposed off at a common treatment, storage and disposal facility (TSDF). It is the centre where the hazardous waste from waste generators is disposed (in the nearby area).

Treatment of Hazardous waste

(i) Physical Methods

Methods of sedimentation, filtration, adsorption and reverse osmosis are employed to remove waste particles.

(ii) Chemical Methods

Neutralization of strong acids and bases, oxidation of cyanides, reduction of Cr (VI) and alkaline and acid hydrolysis of pesticides are the chemical methods employed for the removal of toxic waste.

(iii) Biological Methods

Decomposers breakdown complex hazardous waste into simpler form. *e.g. Phanerochaete chrysosporium* is reported to break toxic aromatic hydrocarbons to Carbon-di-oxide and water.

Convention of Hazardous waste

Hazardous waste management is a new concept for most of the Asian countries including India is a Party to the Basel Convention on transboundary movement of hazardous wastes. The basic objectives of the Basel Convention are control and reduction of transboundary movements of hazardous and other wastes subject to the Basel Convention, prevention and minimization of their generation.

7.1.9 Noise Pollution

Unwanted sound which negatively affects living systems is noise. Noise pollution is also known as sound pollution.

Measurement of Sound

Noise measured in Decibels, as a measure of sound intensity level or sound pressure level.

$$dB = 10\log 10\left\{\frac{I_m}{I_0}\right\}$$

where, I_m is intensity measured and I_0 is reference intensity.

Table 7.9: list of Hazardous Wastes or Chemicals
(i) air pollutants and adverse effects on green plants

Toxicants/ Pollutants	*Source of Contamination*	*Health Effect (Long-term exposures causes death)*
Pesticides	(*a*) Modern agriculture (*b*) Leachate from landfill sites	(*a*) Biomagnification of pesticides leads to reproductive and endocrinal damage (*b*) Neurotoxins
Synthetic organics	Industrial chemicals and agricultural pesticides.	Aromatic compounds cause cancer.
Lead.	Pipes, fittings, solder, and the service connections of house-hold plumbings contaminating drinking water.	Lead is neurotoxin, adversely affecting children and pregnant women.
Fluoride.	Present in rocks of some areas *e.g.* feldspar, contaminating drinking water.	Fluorosis (weakening of teeth and bones)
Arsenic	(i) Arsenic occurs naturally (ii) contamination in aquifers (iii) by phosphorus from fertilizers.	Arsenic poisoning.
Heavy metals	Industrial waste, mining waste	(i) biomagnifications (ii) neurotoxins (iii) nephrotoxins
Electronic waste	The ash of discarded part of electronics contains chlorinated and brominated compounds aromatic compounds.	Diseases of nervous system, the endocrine system and reproductive functions.
Air pollutants		
Aldehydes	Thermal decomposition of fats, oils, glycerol	Nasal and respiratory tracts irritation
Ammonia	Chemical process – dye making explosives fertilizers	Inflame upper respiratory passage
Arsenic	Coal and oil furnaces glass manufacturing	Arsenecosis: Damage kidneys, lungs and skin cancer.
Benzene	Refineries, motor vehicles	Leukemia (blood cancer)
Cadmium	Coal and oil furnaces	Nephrotoxin: Kidneys damage
CO	Gasoline motor exhaust	Damages lungs, carboxyhaemoglobin formation, heart attack
Chlorine	Chemical industries	Lung diseases
Nickel	Oil and coal furnaces	Lung cancer
Nitrogen oxides	Motor Vehicles exhausts	Brohchitis and other lung diseases.
Ozone	Secondary pollutant, photochemical smog	Irritales eyes, aggravate asthma.
Lead	Motor Vehicles exhausts	Brain damage, affects growth high B.P.
SO_2	Coal and Oil combustion	Eye irritation and lung diseases.
Suspended solids	Manufacturing process incinerator	Eye irritation and lung disases

Sources

1. Automobiles
2. Industrial Machines
3. Defense equipments
4. Construction Activities
5. Loudspeaker
6. Mobile phones

Effects

Noise adversely affect the human and animal health.

Pschychological Effects

1. It causes headache, nausea, vomiting.
2. It reduces concentration of mind.

Physiological Effects

1. It adversely affect heart functioning.
2. Noise disturbs sleep, may lead to loss of sleep known as insomnia.
3. It adversely effects communication.
4. It may cause permanent damage of ears.

Control

1. Use of ear plugs.
2. Use of silencers in vehicles
3. Proper servicing of vehicles
4. **Acoustic zoning:** Homes, hospitals and schools should be made far from industrial and traffic (high noisy) areas, known as acoustic zoning.
5. **Bioremediation**: Certain plants absorb the noise waves known as bioremediation *e.g.* hedge plants.

7.2 Air-borne Diseases

An **airborne disease** is any disease that is caused by pathogens and transmitted through the air. The relevant pathogens may be viruses, bacteria, or fungi, and they may be spread through coughing, sneezing, raising of dust, spraying of liquids. Airborne diseases do not include conditions caused simply by air pollution such as dusts and poisons. Air pollution plays a direct role in enhancing the adverse effects of air-borne diseases.

Inflammation in the nose, throat, sinuses and the lungs is caused due to airborne pathogens or allergens. They affect a person's respiratory system or even the rest of the body. Anthrax (inhalational), Chickenpox, Influenza, Measles, Smallpox and Tuberculosis are air-borne diseases. Prevention of airborne diseases include washing hands, using appropriate hand disinfection, getting regular immunizations against diseases believed to be locally present, wearing a respirator and reducing time spent in the presence of any patient.

7.3 Toxic Substances: Toxicant, Toxicity and Toxicology

Substances those cause toxicity are known as toxicants. According to "Paracelsus" every substance is toxic if taken in excess. Toxicity is the property of a substance due to which usually an adverse effect is caused on living organisms. A toxicant undergoes biochemical transformations, which results in increase or decrease of toxicity. A toxic response is understood with following steps:

(i) *Metabolism* Metabolic processes and absorption of toxicant or conversion of protoxicant to more toxic form, storage and excretion steps are involved in toxic responses.

(ii) *Modified receptor* Toxicant binds to the receptor site to develop modified receptor.

(iii) *Biochemical response* Variations in temperature, pulse rate, respiratory rate and blood pressure, hallucinations, convulsions, coma, ataxia, paralysis and death are the physiological responses.

Potentially toxic chemicals used in industries and households cause adverse health impacts. The science of study of toxic chemicals on living organisms and other biological systems and the mechanism of causing toxicity is known as **toxicology or environmental toxicology**. The word toxicology is derived from two words: 'toxicum' (or toxin or toxicon) meaning poison and logos means study or knowledge. Thus toxicology is the basic science of poisons. The branch of science which studies the effect of toxic materials in industry is known as **industrial toxicology**. Chemical properties of toxic substances, their biochemical reactions and how xenobiotic substances and their metabolites react biochemically in an organism to cause toxicity are studied under **toxicological chemistry**.

7.3.1 Factors affecting toxicity

Toxicity of a substance *i.e.* toxic manifestations in an organism that is exposed to the substance depend on following factors:

(1) Factors Rrelated to Property of Organisms

(i) Species

A chemical that is toxic to individuals of one species at a given dose may have little effect in the individuals of other species. Individuals of a single species vary greatly in their sensitivity to a chemical *i.e.* some may be sensitive or highly sensitive or others may be resistant.

(ii) Sex

Male and female of the same species and strain are found to react similarly towards toxicants. Differences are observed in their susceptibility towards toxic compounds. Prolonged sleep is caused in female rats as compared to male rats, when treated with many barbiturates. Only male rats on treatment with Chloroform suffered nephrotoxicity.

Exposure to Toxic Substance	Adverse Effect in One Species	Little Effect in Another Species
Silica dust	Silicosis is caused in humans (miners)	Exposed mules did not show adverse effect.
2 Naphthylamine	Bladder tumour in exposed dogs and humans	Exposed rat, rabbit, guinea pig.
Aniline	Metabolized in cat and dog mainly to o-aminophenol (more toxic compound)	Rat and hamsters metabolize to less toxic compound p-aminophenol.
Malathione (insecticide)	Oxidised in insects to malaoxon which then binds to and inhibits enzyme cholinesterase.	Is metabolized by hydrolysis in mammals.
Ethylene glycol	Metabolized to oxalic acid in cat.	Rat and rabbit are less toxic effect.
Chloroform	Nephrotoxic in male mouse	Female mouse.
Many barbiturates	Prolonged sleep in female rats.	Male rats.

A chemical may affect the testes of a male but not the ovaries of a female or viceversa. *E.g.* women can get intoxicated with alcohol more rapidly than a man of same weight because they have less alcohol dehydrogenase in their bodies, which is the first enzyme involved in alcohol metabolism.

(iii) Age

Young one's are highly susceptible (1.5 to 10 times more) to many toxicants than adults, may be due to deficiency of various detoxification enzyme systems and higher rate of absorption. *E.g.* Young children absorb lead (4-5 times) and cadmium (20 times) to a greater extent than adults. Higher susceptibility to morphine in young ones is due to a less efficient **blood-brain barrier**. DDT is reported 20 times more toxic to young ones than to adults. However, most of the CNS stimulants are less toxic to neonates. Old animals and humans are more susceptible to many toxicants.

(iv) Nutritional Status

The principal biotransformation of toxicants is catalyzed by the microsomal enzyme *Mixed Function Oxidase* (MFO) system. The balanced diet content is important for proper functioning of MFO activity. It is depressed by deficiency of essential fatty acids, proteins, vitamin A, vitamin C and vitamin E, causing toxic effects. *E.g.* Vitamin A deficiency increases the adverse effects of carcinogens to the respiratory tract.

(v) Diseases

Diseased persons are highly prone to further toxic effects.

(2) Environmental Factors

Environmental factors play an important role in the process of metabolism of toxicants, thus cause variations in toxic effects.

(i) Physical Factors

Temperature changes may cause variations in toxicity. *E.g. Colchicine* and *Digitalis* are more toxic to the rat as compared to frog. However, it is observed that toxic effects to the frog can be enhanced by increasing temperature. Similarly, at high altitudes the toxicity of Digitalis is reduced and of amphetamine is increased.

(ii) Social Factors

Individuals more friendly in the society with positive attitude towards life, enhances the ability of body to detoxify the toxic substances. Tension has also reported to enhance the toxic effects.

(3) Chemical Interactions

Toxicant causes injury at the site of contact and after being absorbed by the target organ in the body. The nature and intensity of the toxic effect is dependent upon the concentration of the toxicant and duration of the exposure. The toxicity of the administered dose depends upon the metabolism of the toxicant *i.e.* absorption, distribution, binding and excretion.

7.4 Carcinogens

Any substance, radionuclide, or radiation directly involved in causing cancer due to the ability to damage the genome or to the disruption of cellular metabolic processes. Natural carcinogens are Aflatoxin B_1, produced by the fungus *Aspergillus flavus* growing on stored grains, nuts and peanut butter. Certain viruses such as hepatitis B and human papilloma virus have been found to cause cancer in humans. The first virus that caused cancer in animals is Rous sarcoma virus, discovered in 1910 by Peyton Rous. Other infectious organisms which cause cancer in humans include some bacteria (*e.g. Helicobacter pylori*) and helminths (*e.g. Opisthorchis viverrini* and *Clonorchis sinensis*).

Dioxins and dioxin-like compounds, benzene, kepone, EDB, Benzo[*A*]Pyrene, tobacco-specific nitrosamines such as nitrosonornicotine, and reactive aldehydes such as formaldehyde and asbestos are man-made carcinogens. Vinyl chloride, from which PVC is manufactured, is a carcinogen and thus a hazard in PVC production.

Cancer is any disease in which normal cells are damaged and do not undergo scheduled cell death as fast as they divide via mitosis. Carcinogens usually increase the risk of cancer by altering cellular metabolism or damaging DNA directly in cells, which interferes with biological processes, and induces the uncontrolled, malignant division, ultimately leading to the formation of tumors.

Co-carcinogens are chemicals that do not necessarily cause cancer on their own, but promote the activity of other carcinogens in the process of causing cancer.

Participatory Learning

(a) Objective Questions

1. Environmental pollution adversely affects ______.
2. Acid rain is caused due to ______, ______ and ______. (***UPTU, 2007***)
3. Corrosion of Taj Mahal is attributed to ______.
4. PAN is formed by the reaction of oxides of nitrogen and ______ in the presence of sunlight. (***UPTU, MCA 2010***)
5. PAN is mainly______ and is an important ______ pollutant.

True/False

6. All pollutants are degradable. (***UPTU, 2007-8***)
7. Eutrophication of lakes is caused by intake of excess of nutrients in the water-body.

(b) Very Short Answer Type Questions

1. Define environmental pollution.
2. What is the cause of minamata disease and itai-itai disease?
3. What do you understand by smog? Does it causes environmental pollution?

4. What is the role of bioremediation in control of water and marine pollution?
5. Give the expression for measurement of noise. What is the unit of measurement of sound waves?
6. Define the process of eutrophication? (*MTU, 2013-14*)
7. Give the major composition of PAN. Is PAN secondary pollutant?
8. What do you understand by endemic? Explain an endemic disease.
9. How does air pollution affect green pigment of leaf?
10. What is phaeophytin?
11. What is zero-waste management?

(c) Short Answer Type Questions

1. Differentiate between losangeles and London smog.
2. Differentiate between primary and secondary air pollutants.
3. Does excess of nutrient supply cause water pollution? How?
4. Write a short note on Bhopal gas tragedy.
5. Comment "air pollution is a global environmental problem"
6. List the solid waste and hazardous wastes which we encounter in our daily life. (*UGC-NET, 2002*).
7. What are the causes, effects and control measures of acid rain? or Comment "Taj Mahal is corroding".
8. Explain the case study of Chernobyl disaster and Fukushima disaster. (*MTU, 2011*)

(d) Long Answer Type Questions

1. List the major air quality standards given by CPCB. (*UPTU, 2008, 2010*)
2. Explain four major air pollutants and their consequences. (*UPTU, 2007*)
3. What is air pollution? What are its effects on human health? (*UPTU, 2005, 2008*)
4. Discuss the measures used for controlling air pollution. (*UPTU, 2005*)
5. What are 'solid wastes'? Discuss their types, effects and name the various methods used to dispose solid wastes. Explain any one of them with merits and demerits. (*UPTU, 2005, 2006*) (*MTU, 2013-14*)

6. Describe the idea of solid waste management. Distinguish between the incineration and combustion methods of disposal of solid wastes. (***UPTU, 2007***)
7. Describe solid waste. What are the different types of solid waste? Discuss its sources and effects. (***UPTU, 2008, 2010***)
8. Define environmental pollution. Discuss the requirement of a non-polluted environment. (***UPTU, 2007, 2009***)
9. How can you as an individual prevent environmental pollution? Why such an effort is important at an individual level? (***UPTU, 2006***)
10. Discuss water treatment with layout of a water treatment plant. (***UPTU, 2006***) (***MTU, 2013-14***)
11. What are the causes, effects and control measures of thermal pollution. (***UPTU, 2010***)
12. Enumerate various effects of soil pollution. How do industrial, agroproducts and pesticides deteriorate the soil? (***UPTU, 2009***)
13. Define land pollution. Discuss the causes of land pollution and their control. (***UPTU, 2008***)

Answers to Objective Questions

1. Human health
2. Oxides of sulphur, nitrogen and carbon
3. Acid rain
4. Hydrocarbon
5. Per oxyAcetyl Nitrate, secondary
6. False
7. True

Chapter 8

Environmental Impact Assessment

8.1 Environmental Impact Assessment (EIA)

Environmental Impact Assessment is a continuous process, in which proponent of a mega-developmental project analyses its positive and negative environmental effects. It minimizes negative environmental effects of the mega-developmental project. Thus, helps in Sustainable Development (It is a tool to sustainable Development).

IAIA (International Association of Impact Assessment) defines EIA as "the process of identifying, predicting evaluating and mitigating the biophysical, social and other relevant effects of development proposals prior to major decisions being taken and commitments made.

8.2 Government Body which Executes EIA in India

Ministry of Environment and Forests (MoEF) is responsible for EIA. The programme of EIA was taken up by the government in 1977-78 when the Planning Commission suggested that all the river valley projects should be appraised from environmental point of views. In 1992, after consultation with all the ministries concerned, the MoEF issued a draft notification making environmental impacts assessment statutory for selected activities. The Ministry of Environment and Forests (MOEF), Government of India under the Environment (Protection) Act, 1986 promulgated a notification **The Environment Impact Assessment Notification, 1994**

8.3 Environmental Effects Analysed under EIA

The following environmental effects are analysed in EIA:

1. Social and economic effects are analysed.
2. Air pollution
3. Noise pollution

4. Adverse impact on land fertility *i.e.* soil erosion.
5. Surface water and ground water pollution.

8.4 Process of EIA

The complete process of EIA is mentioned in **Figure 8.1**. The important steps are screening (selection of the megadevelopmental project to be executed for EIA) and scoping (local people have right to express their comments to the government regarding the project's impact on their life) The documents prepared showing systematic possible effects of mega-developmental activities during the exploitation and processing of natural resources on natural environment is called as **Environmental Impact Statement.** Negative environmental impacts are minimized, helping in final decision making and inspections.

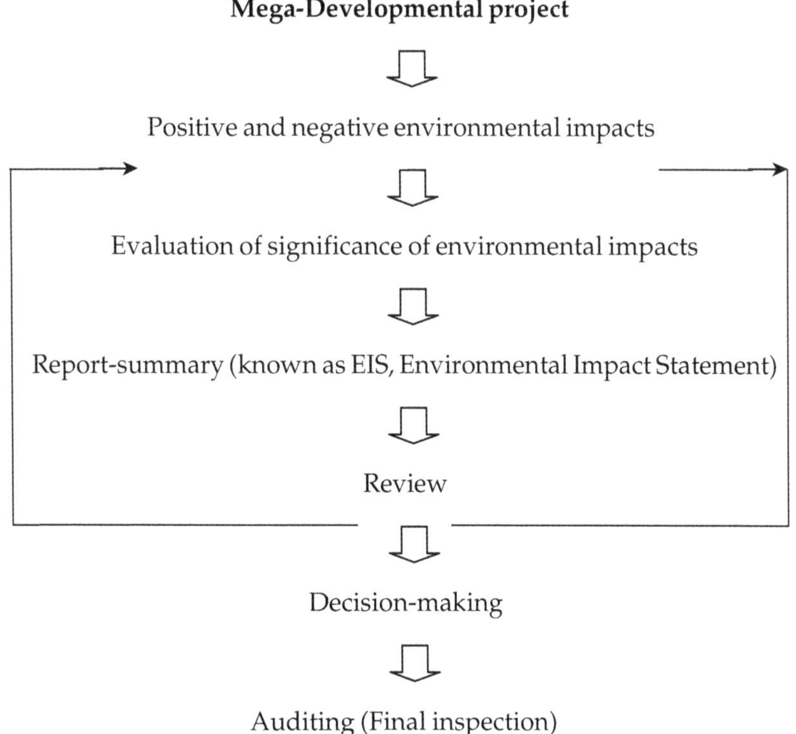

Figure 8.1: The process of Environmental Impact Assessment.

8.5 EIA Ruling 1984

Objectives: **Ministry of Environment and Forest Notification** on 27 January 1994 discusses the list of **Environment Impact Assessment of developmental projects.**

Schedule -I

List of Projects Requiring Environmental Clearance from the Central Government

1. Nuclear Power and related projects such as heavy eater plants. Nuclear fuel complex and rare earth.
2. River Valley projects including hydel power, major irrigation and their combination including flood control.
3. Ports, Harbours, Airports (except minor ports and harbours).
4. Petroleum Refineries including crude and product pipelines.
5. Chemical Fertilizers (Nitrogenous and Phosphatic other than single superhosphate).
6. Pesticides (Technical).
7. Petrochemical complexes (both Olefinic and Aromatic) and Petrochemical intermediates such as DMT, Caprolactam, LAB etc. and production of basics plastics such as LDPE, HDPE, PP, PVC.
8. Bulk drugs and pharmaceuticals.
9. Exploration for oil and gas and their production, transportation and storage.
10. Synthetic Rubber.
11. Asbestos and Asbestos products.
12. Hydrocyanic acid and its derivatives.
13. (a) Primary metallurgical industries (such as production of Iron and Steel, Aluminium, Copper, Zinc, Lead and Ferro Alloys).

    (b) Electric arc furnaces (Mini Steel Plants).
14. Chlor – alkali industry.
15. Integrated paint complex including manufacture of resins and basis raw materials required in the manufacture of paints.
16. Viscose Staple fibre and filament yarn.
17. Storage batteries integrated with manufacture of oxides of lead and lead antimony alloy.
18. All tourism projects between 200-500 meters of High Tide Line or at locations with an elevation of more than 1000 meters with investment of more than Rs. 5 crores.
19. Thermal Power plants.
20. Mining projects (major minerals) with leases more than 5 hectares.
21. Highways Projects except projects relation to improvement work including widening and strengthening of roads with marginal land acquisition laong

the existing alignments provided it does not pass through ecologically sensitive areas such as National Parks, Sanctuaries, Tiger reserves, Reserve forests.

22. Tarred Roads in Himalayas and/or Forest areas.
23. Distilleries.
24. Raw Skins and Hides
25. Pulp, paper and newsprint.
26. Dyes.
27. Cement.
28. Foundries (individual)
29. Electroplating.

Schdule II and III has application form for developmental project and its report to be submitted by the chairman.

Participatory Learning

(a) Objective Questions

1. Environmental Impact Assessment is a ______ process in which ______ and ______ environmental effects are analyzed.
2. EIA helps in ______.

(b) Very Short Answer Type Questions

1. Define screening.
2. How EIA helps in sustainable development?

(c) Short Answer Type Questions

1. What is Environmental Impact Statement? Give its importance.
2. Differentiate between screening and scoping.

(d) Long Answer Type Questions

1. Define EIA. Explain the process or stages of EIA. (*UPTU, 2009; MTU, 2012*)
2. Comment "Environment and development run hand-in-hand". Or " EIA is a tool to sustainable development". Or Explain the functions of EIA. (*MTU, 2013-14*)
3. Explain the role of EIA in river valley projects.

Answers to Objective Questions

1. Continuous, positive, negative
2. Sustainable development

Chapter 9

Environmental Laws

9.1 Environmental Laws: Provisions in the Indian Constitution towards Environmental Protection

Environmental concerns were incorporated into the Indian constitution (act of 1976) through the 42nd amendment, after the Stockholm conference, with specific provisions towards environmental conservation and human rights (**Table 9.1**).

Table 9.1: Environmental concerns in the constitution of India

Constitution of India	*Description of Environmental Conservation*
Article 21 (Right to Life)	**"No person shall be deprived of his life or personal liberty except according to procedure established by law"**.
Article 14 (Right to Equality)	**"The State shall not deny to any person equality before the law or the equal protection of the laws within the territory of India"**
Article 48 A (Directive Principles of State Policy)	The state shall endeavour to protect and improve the environment and safeguard the forests and wildlife of the country.
Article 51A (g)	Responsibility on every citizen to protect and improve the natural environment including forests, lakes, rivers and wildlife, and to have compassion for living creatures. Thus, protection of natural environment and compassion for living creatures was made the positive fundamental duty of every citizen

9.1.1 Salient Features of Air (Prevention and Control of Pollution) Act, 1981*

Objective

(a) **Prevention and control of air pollution, in whole India** : Industries should treat their gaseous waste to ensure control of air pollution.

(b) **Natural resource conservation:** Maintaining the quality of air to ensure good health of the nearby inhabitants.

Definitions

(a) Air Pollution

Presence of any gas beyond the threshold level in the atmosphere (*i.e.* air pollutant), which adversely affect the human health, is known as air pollutant.

(b) Air Pollutant

Any substance: solid, liquid or gaseous beyond the threshold level in the atmosphere (*i.e.* air pollutant), which adversely affect the human health, is known as air pollutant.

Environmental Provisions

(a) Construction of Central Pollution Control Board

Functions of Central Pollution Control Board

(i) Power to declare air pollution control areas.

(ii) Power to establish standards for emission of air pollutants from automobiles.

(iii) Power to restrict the use of certain industrial plants

(iv) Power of entry and inspection in the industries at any point of time.

(v) Power to take air samples to analysis of the pollutants present.

(vi) Advise central government regarding air pollution related issues.

(vii) Co-ordinate the State Pollution Control Boards and resolve disputes between them.

(viii) Provide technical assistance and guidance to state boards.

(b) Construction of State Pollution Control Board

SPCB advise the state government regarding following issues of air pollution:

(i) Under the section 19 of the Act, power to declare air pollution control areas.

(ii) Power to establish standards for emission of air pollutants from automobiles.

(iii) Under the section 21 (1), (2) and (3), power to restrict the use, establishment or operation of certain industrial plants.

(iv) Power of entry and inspection in the industries at any point of time.

(v) Power to take air samples to analysis of the pollutants present.

(vi) Under the section 19.1, prohibition of burning or use of any appliance or fuel of any air pollution causing material, in air pollution control areas.

Punishment and penalties

The person or organization that does not follow the law shall be given following punishments:

Punishment (i) or (ii) or both	*Description of Punishment*
Imprisonment (under section 22, 22 (A))	Three months
Fine	10,000 rupees

9.1.1.1 Amendments

☆ **The Act was amended in 1987.**

The objective was empowerment of central and state pollution boards to handle emergencies and recover the penalties from the offenders.

☆ **The Air (Prevention and Control of Pollution) Rules, 1982.**

Defined the procedures for conducting meetings of the boards, the powers of the presiding officers and decision-making,

Motor Vehicles Act, 1988 was constructed to regulate vehicular traffic and proper packaging, labelling and transportation of the hazardous wastes.

9.1.1.2 Concluding Remarks of Air (Prevention and Control of Pollution) Act, 1981

(i) **Addressing the environmental problem of air pollution**: Ambient air quality standards were established, under the 1981 Act.

(ii) **Control or abatement of air pollution**: Prohibiting the use of polluting fuels and substances and appliances in air pollution control areas. Under the Act establishing or operating of any industrial plant in the pollution control area requires consent from state boards. The boards are also expected to test the air in air pollution control areas, inspect pollution control equipment, and manufacturing processes.

9.1.2 Salient Features of Water (Prevention and Control of Pollution) Act, 1974*

Objective

(a) **Prevention and control of water pollution**: Industries should treat their waste to ensure control of water pollution.

(b) **Natural resource conservation:** Maintaining the quality of water to ensure good health of the nearby inhabitants.

Definitions

Water Pollution

Presence of any substance beyond the threshold level in the hydrosphere, which adversely affect the human health, is known as water pollution.

Water Pollutant

Any substance: solid, liquid or gaseous beyond the threshold level in the hydrosphere (*i.e.* water pollutant), which adversely affect the human health, is known as water pollutant.

Environmental Provisions

Construction of Central Pollution Control Board

Functions of Central Pollution Control Board

(i) Power to declare water pollution control areas.

(ii) Power to establish standards for emission of water pollutants from industries.

(iii) Power to restrict the use of certain industrial plants

(iv) Power of entry and inspection in the industries at any point of time.

(v) Power to take water samples to analysis of the pollutants present.

(vi) Advise central government regarding water pollution related issues.

(vii) Co-ordinate the State Pollution Control Boards and resolve disputes between them.

(viii) Provide technical assistance and guidance to state boards.

(viii) Construction of **State Pollution Control Board**

Functions of State Pollution Control Board

(i) Power to declare water pollution control areas.

(ii) Power to establish standards for emission of water pollutants from automobiles.

(iii) Power to restrict the use of certain industrial plants.

(iv)

Power of entry and inspection in the industries at any point of time.

(v) Power to take water samples to analysis of the pollutants present.

(vi) Advise the state government regarding causes, effects and control measures of water pollution.

Punishment and Penalties

The person or organization that does not follow the law shall be given following punishments:

Punishment (i) or (ii) or both	*Description of Punishment*
Imprisonment	Three months
Fine	10,000 rupees

9.1.2.1 Concluding Remarks of Water (Prevention and Control of Pollution) Act, 1974

(i) **Addressing the environmental problem of air pollution**: Ambient water quality standards were established, under the 1974 Act.

(ii) **Control or abatement of water pollution**:

Prohibiting the use of polluting fuels and substances and appliances in water pollution control areas.

Establishing or operating of any industrial plant in the pollution control area requires consent from state boards.

The state boards have to test the water in water pollution control areas and pollution control equipment and manufacturing processes.

9.1.3 Salient features of Forest Conservation Act, 1980*

Objectives

(i) Conservation of forest.

(ii) Maintaining harmony between man and biodiversity.

(iii) Efficient use of timber and non-timber forest products to ensure their sustainable supply.

(iv) Restricting the use of forest land for non-forest purpose.

Definitions

Forest is the biotic community of trees, shrubs, herbs and grasses.

Forest= Trees + Shurbs + Herbs + Grasses

Salient Features

1. Re-afforestation or compensatory afforestation

Punishment and Penalties

9.1.3.1 Amendments in Forest Conservation Act, 1980 (Carryover, MTU, 2012-13)

2. Indian Forest Act, 1865, Amendment 1878. In 1865, the first Indian Forest Act was passed. It was amended in 1878 when a comprehensive law, The Indian Forest Law Act VII, came into force. The provision of this Act established a virtual State monopoly over the forests in a legal sense on one hand, and attempted to establish, on the other, that the customary use of the forests by the villagers was not a 'right', but a 'privilege' that could be withdrawn at will.

Indian Forest Act, 1927. The Act consolidated and reserved the areas having forest cover or significant wildlife and defined the procedure for declaring forest as reserved, protected and village forest and also levying penalties for not following the provisions of the Act.The Forest (Conservation) Rules were made 1981 in order to exercise the powers conferred by Section 4 of the Act.

---

* Natural resource legislative acts.

9.1.4 Salient Features of Wildlife Protection Act, 1972*

Objectives

(i) **Natural resource conservation**: Protection and conservation of biodiversity to maintain balance of essential ecological harmony.

(ii) Maintaining harmony between man and biodiversity.

(iii) **Sustainability**: Efficient use of natural resources obtained by biodiversity to ensure their sustainable supply.

Definitions

Wildlife is the biotic community of variety of animal and plant species found in the forest region.

Wildlife= biotic component of forest.

Salient Features

Constitution of Wild-life Advisory Board

Wildlife advisory board aims at biodiversity conservation and follows following duties:

(i) Identifying areas to be declared as national parks and sanctuaries.

(ii) Identifying the species which needs conservation efforts.

(iii) Developing a strategy to conserve biodiversity.

Punishment and Penalties

The person or organisation that does not follow the law shall be give following punishments:

	Punishment (i) or (ii) or both	Description
(i)	Imprisonment	Three Years
(ii)	Fine	Twenty five thousand rupees

9.1.4.1 The Wildlife (Protection) Act, 1972, Amendment 1991.

The complete ban on hunting was made more effective by the Amendment Act of 1991.

9.1.5 Salient Features of Environment (Protection) Act, 1986

Objectives

(i) **Natural resource conservation:** Protection of our environment.

(ii) **Disaster management**: Prevention of hazards to all living things (humans, plants and animals) and property.

(iii) Maintaining harmony between human beings and environment.

Definitions

Environment is the biotic community of trees, shrubs, herbs and grasses.

☆ **Environmental pollution**

☆ It gives specific rules (regulations) for

(i) Hazardous Waste.

(ii) Hazardous Chemicals.

(iii) Biomedical Waste.

(iv) Genetically Engineered Organism (GEO).

(v) Environmental Impact Assessment.

☆ **Important terms used in this act**

Environment: It defines 'environment' as interrelationship among air, waste and land and human beings, other living creatures, plants microorganisms and property.

Environmental pollutant means 'any solid, liquid or gaseous substance present in such concentrations as may be injurious to environment'.

Environmental pollution: means 'the presence of any environmental pollutant in the environment'.

Hazardous substances: means 'any substance which is liable to cause harm to human beings, other living beings and property by reason of its chemical or physicochemical properties of handling'.

Salient Features

(i) This Act is known as **umbrella legislation** because it provides environmental protection in a sound legal framework (**Figure 9.1**).

(ii) It is most detailed or comprehensive law.

(iii) It is most stringent (strict) law. *i.e.* Under the section 3 and 6 of this Act, the central government is empowered to take strict necessary steps to protect and improve the quality of the environment and regulate environment pollution.

(iv) Thus, it is most important law.

(v) Construction of **Central Pollution Control Board** is mentioned.

Appointment of Chairperson and other Members

Functions of Central Pollution Control Board

(i) Make rules for environmental protection.

(ii) Standards of air, water and soil quality.

(iii) Maximum concentration of pollutants.

(iv) Procedure and safeguards for handling hazardous waste.

(v) Environmental Impact Assessment certification (list of 29 types of projects is given in appendix).

State Pollution Control Board

Industrial projects with investments above Rs 500 million must obtain clearance from MoEF and an NOC (No Objection Certificate) from the SPCB and the State Forest Department if the location involves forestland.

Handling and Management of Hazardous Waste

Under the Section 8, procedure for handling hazardous substances has been explained.

Automobile Pollution

Various aspects of vehicular pollution have also been notified under the EPA of 1986. Mass emission standards were notified in 1990, which were made more stringent in 1996 **(Table 9.2)**.

Table 9.2: Air Pollution Emission Standards.

Air Pollution Emission Standards	*EPA Notification*
Euro I* (for car manufacturers)	May 1999
Euro II* norms (for new non-commercial vehicle sold in Delhi)	April 2000
Euro IV	Notified in 11 cities including NCR since April 2010.

* Stringent Euro I and II emission norms were notified by the Supreme Court on April 29, 1999 for the city of Delhi.

Guidelines for Environmental Protection

EPA keeps providing guidelines for environmental protection as and when required. A list of guidelines provided by EPA is given in **Table 9.3**.

Table 9.3: Examples of EPA Notifications.

EPA Notification	*Details of Notification*
Doon Valley Notification (1989)	**Ecologically-sensitive areas** Notifications under the EPA for the protection of ecologically-sensitive areas *e.g.* prohibits the setting up of an industry with daily consumption of coal/fuel is more than 24 MT (million tonnes) in the Doon Valley,
Taj Trapezium Notification (1998)	No power plant could be set up within the geographical limit of the Taj Trapezium.
Disposal of Fly Ash Notification (1999)	No person within a radius of 50 km from a coal-or lignite-based power plant shall manufacture clay bricks or tiles without mixing at least 25 per cent of ash with soil on a weight-to-weight basis.

Punishment and Penalties

The person or organization that does not follow the law shall be given following punishments (amendments are done time to time):

Punishment (i) or (ii) or both	Description of Punishment
(i) Imprisonment	In case the violation of law continues beyond 1 year, the offender shall be punishable imprisonment of 7 years.
	Imprisonment of 5 years if the person violates the law.
	Criminal liability is fixed on directors and principal officers, if the law is violated by industry.
(ii) Fine	1 lakh rupees

9.1.5.1 Amendments in EPA

The Environmental Impact Assessment of mega-development projects notification, (1994 and as amended in 1997, revised in 2006) are:

(i) developmental projects listed under Schedule I require environmental clearance from the MoEF.

(ii) All developmental projects whether or not under the Schedule I, if located in ecologically sensitive areas must obtain MoEF clearance.

(iii) Projects under the delicenced category of the New Industrial Policy also require clearance from the MoEF.

EPA (Section 3 and 6)

Strict measures of control of environmental pollution and environmental protection.

(Standards for emissions and discharges; regulating the location of industries; management of hazardous wastes, and protection of public health and welfare.)

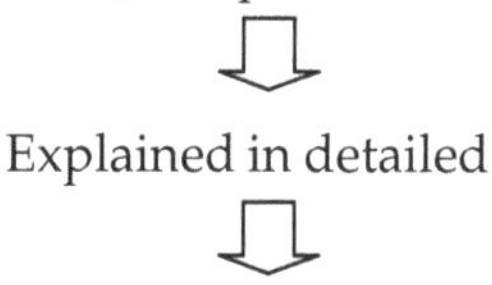

Explained in detailed

Thus, most important law and umbrella legislation.

Figure 9.1: Conceptual framework of Environmental Protection Act, 1986.

9.2 Role of Government in Environmental Protection

(a) **Environmental laws** played important role in conservation of natural resources.

(b) **Provision of Funds** for environmental conservation *e.g.* Department of Biotechnology provides funds for environmental research. Ministry of Environment and Forests help innovation environmental conservation projects. Central Pollution Control Boards also help in environmental protection

(c) Role of government agencies in environmental protection

Ministry of Environment and Forests is the government agency responsible for environmental protection. The National Council for Environmental Policy and Planning within the Department of Science and Technology evolved the Ministry of Environment and Forests (MoEF) in 1985.

Botanical Survey of India (BSI): The Botanical Survey of India (BSI) was established in 1890 at the Royal Botanic Gardens, Calcutta with different circle offices to establish a data of plants that require conservation efforts.

Table 9.4: National policies towards environmental conservation

National Forest Policy, 1988 (*UPTU, 2007*)	**Forest conservation**: stops soil erosion; livelihood security to the local people depending on forest, 66 per cent forest area is recommended; reduces the risks of floods, landslides and drought.
National Water Policy, 2002	**Planning, development and management of water resources**: Water conservation, Adequate safe drinking water supply. Environmental planning and management of water resources projects. Water zoning of the country and the development activities are guided and regulated in accordance with availability and constraints of water resources. **Disaster management:** (floods and drought) predicting, prevention, mitigation, relief and rehabilitation measures.
National Agricultural Policy	To attain an annual growth rate of 4 per cent in the agricultural sector over two decades (2000-2020). The new Agriculture Policy, 2000, of India emphasized on Sustainable agriculture, Food and nutritional security, Technology generation and transfer, risk management and organizational framework.
The National Environmental Policy, 2006	Integrated approach of environmental management: role of EIA; management of coastal zones, wetlands and river systems; of mountain ecosystems; land use planning; watershed management and disaster management.
Urban Sanitation Policy, 2008	Water and garbage related health disasters (epidemics), solid waste management; generation of industrial and other specialized/hazardous wastes; drainage; as also the management of drinking water supply, for a city-sanitation plan and its implementation.
National Disaster Management Policy 2009	Definition of Disaster (as per the DM Act of India) is damage to environment. Factors that increase people's vulnerability to a disaster is environmental degradation and climate change.

Participatory Learning

(a) Objective Questions

1. The most stringent environmental law is ______. (*UPTU, 2008*).
2. The law which is not applicable to Jammu and Kashmir region is ______. (*UPTU, 2009*)

(b) Very Short Answer Type Questions

1. Which articles in our constitution state the importance of provisions of environmental protection?

2. When did we enforce environment protection?
3. Which environmental law is not applicable to all over India?

(c) Short Answer Type Questions

1. Which environmental law is most stringent? (***UPTU, 2007)*** Why it is called umbrella legislation?
2. Describe the role of government agencies in environmental protection (***UPTU, 2011***)

(d) Long Answer Type Questions

1. Describe the laws relating to conservation of air, water and environment. (***MTU, 2012)***
2. Describe the salient features of Wild-life protection act, 1972 and its amendments. (***MTU, 2012; 2013-14***).
3. Describe the salient features of Forest Conservation act, 1980 and explain its amendments. (***MTU Carryover, 2013***).
4. Describe the different policies on environmental conservation.
5. List the important notifications for environmental protection under EPA, 1986.

Answers to Objective Questions

1. Environmental Protection Act, 1986
2. Wild-life protection Act, 1972

Chapter 10

Global Environmental Issues

10.1 Green House Effect and Global Warming

Green House Effect

The process shown in the diagram as 1, 2, 3 and 4, due to which large amount of green house gases absorb heat rays is known as Green House Effect, which causes drastic rise in the temperature of earth known as Global Warming (**Figure 10.1**).

Global Warming

The average temperature of the earth is rising every year. The drastic rise in earth's temperature is known as Global Warming. The continuous increase in earth's temperature due to industrialization.

Effects of Global Warming

(i) The melting of polar ice due to high temperature in mountains.

(ii) Rise in sea-level will occur.

(iii) Submergence of areas near sea (coastal areas).

(iv) The animals which live in extreme low temperature will get extinct (finished). *e.g.* Polar Bears, Penguins. The current phase is known as **sixth mass extinction**.

(v) Wheat crop cannot survive at high temperature. This will cause a turning point in our eating habits. Maize will be the future dominant crop because it is able to withstand high temperatures.

(vi) Winters are getting shorter and summers are getting longer.

(vii) 1,50,000 (1 lakh 50 thousand people dying every year: WHO (World Health Organisation) estimates.)

1. represents incoming sun rays to earth's atmosphere.
2. represents some of the solar radiations are reflected back
3. represents large amount of sun-rays are absorbed by Green House Gases.
4. represents large amount of sun-rays are transmitted or reflected to earth's atmosphere.

* represents Green House Gases (GHG).

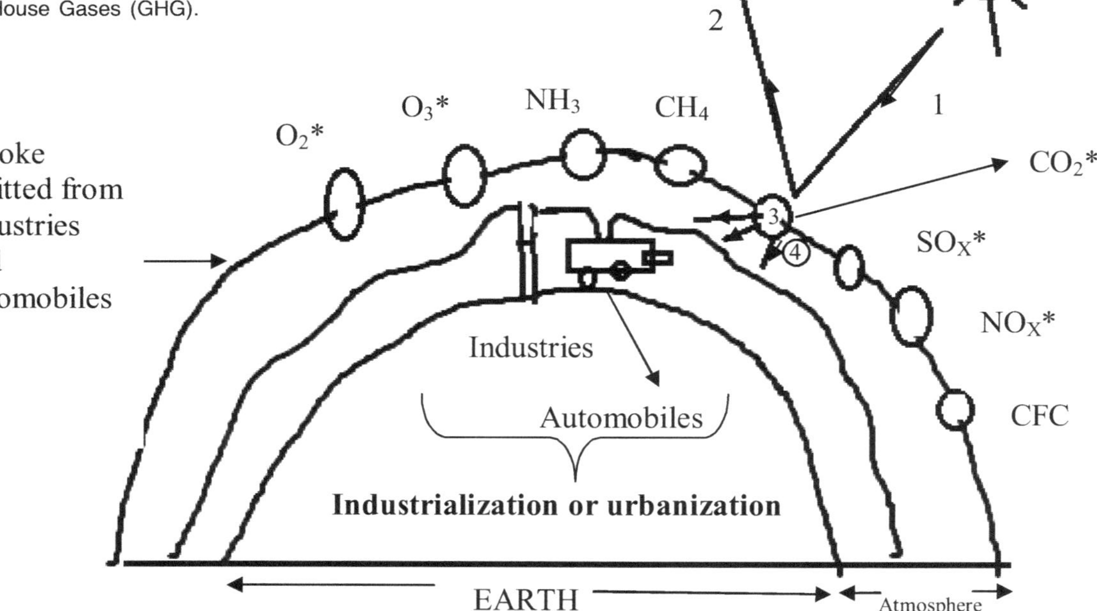

Figure 10.1: Due to Industrialization, large amount of Gases are Released in the Earth's Atmosphere, Leading to Green House Effect.

(viii) Rise in temperature observed in the forest ecosystem made **forest fires** more frequent.

(ix) Diseases *e.g.* Dengue, malaria have also become very common.

(x) **Disturbance in marine and freshwater systems** (associated with rising water temperatures, as well as related changes in ice cover, salinity, oxygen levels and circulation)

(xi) Decrease in algae, planktons and zooplankton and fish number occurs.

The oceans are becoming more acidic due to accumulation of anthropogenic carbon with an average decrease in pH of 0.1 units.

10.2 Global Climate Change

The above discussed effects of global warming when put collectively are known as global climate change.

Climate Change

Climate is the general weather conditions over a period of 25 years. The climatic conditions are found changing. Summers are getting longer and winters are getting shorter.

Impacts of Climate Change: IPCC IV Assessment (2007)

Results of IPCC work on observed and ongoing/projected impacts on climate-change on natural and human environment has been given in its IV assessment report released in 2007 December.

Effects on Aquatic Ecosystem

Melting of ice discharge in many glacier- and snow-fed rivers.

- ☆ Warming of lakes and rivers in many regions, adverse effects on thermal equilibrium and water quality of aquatic ecosystem.

Effects on Terrestrial Biological Systems

- ☆ Earlier timing of spring events, such as leaf-unfolding, bird migration and egg-laying
- ☆ Pole ward and upward shifts in ranges in plant and animal species.

Observed Changes in Climate and Projected Future Trends for India

- ☆ Monsoon rainfall is increasing in the areas found along the west coast, north Andhra Pradesh and north-west India, and decreasing rainfall is occurring over east Madhya Pradesh and adjoining areas, north-east India and parts of Gujarat and Kerala.
- ☆ Increase in heavy rainfall events in Andhra Pradesh, Orissa and Chattisgarh and Madhya Pradesh.
- ☆ Mean annual surface air temperatures show a significant warming of about 0.5 C/100 year during the last century.

Sea level rise along Indian coastline.

The occurrence of severe cyclonic events is projected to increase along the east coast of India, which along with sea level rise will cause increased flood risks.

☆ Glaciers in Himalayas are decreasing at a rapid pace.

Future Scenarios of Change

It is estimateded that rainfall will increase by the end of the 21st century by 15-40 per cent, and the mean annual temperature will increase by 3°C to 6°C, widespread over the country.

Climate-change and Disaster Risks

Vulnerable to cyclones, floods, landslides, drought, earthquakes as well as other localized hazards. drought, floods, heat and cold waves, desertification and coastal hazards like cyclone, coastal and sea erosion, storm surges and flooding.

10.2.1 International Efforts to Control Global Warming or Global Climate Change

Kyoto Protocol (in 1997)

It is a legal agreement in IPCC (Intergovernmental Panel on Climate Change), which emphasizd on following control measures of global warming:

(i) To reduce GHG (Green House Gases) emissions.

(ii) Promoting ecofreindly technologies.

(iii) Encouraging research to address climate change issues (science, impacts and response strategies *i.e.* mitigation).

10.3 Ozone Layer Depletion or Ozone Hole

Thinning of ozone layer in the stratosphere due to excessive use of ozone depleting substances (ODS, in variety of uses: aerosols, cosmetics, rocket propellants, refrigerators and air conditioners as coolants *etc.*). Ozone layer in stratosphere protects harmful UV radiations to come down to the earth surface.

Formation of Ozone

$$O_2 + h\nu \rightarrow O^{\bullet} + O^{\bullet} \text{ (Light) \{Photochemical Reactions)}$$

$$O_2 + O^{\bullet} \rightarrow O_3 + M \text{ (M-catalyst)}$$

Ozone Destruction

Chlorofluoro carbons are known as ODS, emitted into the atmosphere from aerosols, refrigerators, cosmetics, jet-aeroplanes *etc.* Chlorofluoro carbons (CFC) are highly reactive and continuously generate free radicals of chlorine and fluorine, causing ozone depletion reaction as a **chain reaction mechanism**.

$$CF_2Cl_2 + h\nu \rightarrow Cl^{\bullet} + CF_2Cl^{\bullet}$$

$$Cl^{\bullet} + O_3 + h\nu \rightarrow ClO^{\bullet} + O_2$$

$$ClO + O \rightarrow Cl^{\bullet} + O_2$$

$$Cl^{\bullet} + O_3 \rightarrow ClO^{\bullet} + O_2$$

☆ Ozone is a secondary pollutant is troposphere and friend in stratosphere bcause it absorbs harmful UV rays from sun and protects the earth.

10.4 Acid Rain

The precipitation of Acids along with rain causing environmental pollution is known as Acid Rain.

Causes

☆ The smoke emitted from factories and vehicles contains SO_2, NO, NO_2 and small particles. These pollutants react with rainwater to form acids

$$SO_2 + H_2O \rightarrow H_2SO_4$$

$$NO_2 + H_2O \rightarrow HNO_3$$

Effects

☆ The green chlorophyll degrades to brown pigment in the presence of SO_2.

☆ Humans Suffer from Respiratory problems (Bronchitis, Asthma and Cancer).

The exposure of acids to buildings causes corrosion.

Control Measures

Planting air pollution tolerant trees near industries is known as Green-Belt Designing.

10.5 *El Nino*

The El Nino is a ocean warming phenomenon. It is a Spanish word meaning Christ Child. Fisherman along the coast of Eucador and Peru refer to warm off shore ocean conditions occurred near Chritsmas and lasted for several years. It is a meteorological phenomenon wherein the changing higher water temperature and lower atmospheric pressure is observed, due to which flooding occurred in South California and droughts in Australia, India and South Africa in October, 1982. The phenomenon of ElNino is taken together with Global Warming. The temperature of 38C in Texas for 30 days has been observed in 2001.

10.6 *La Nina*

La Nina is commonly known as little sister of El Nino because it is a counter ocean current which arises after El Nino effect finishes in the coastal region of eastern pacific. It causes heavy rains. Severe floods in Australia in 1999 occurred due to La Nina.

10.7 Automobile Pollution

Causes

(*i*) Vehicles are the largest source of air pollution.

(*ii*) Due to incomplete combustion of fuel in the combustion chamber of the engines of vehicles, Carbon monoxide, Oxides of Nitrogen, lead, Particles are released to atmosphere.

Table 10.1: Environmental Effects of air pollutants

Pollutants	*Effects on Environment*
CO	It reduces Oxygen Carrying capacity of blood by forming Carboxyhaemoglobin. (***UPTU 2007-08***)
NO	Respiratory problems.
SO_2	Respiratory diseases.
Acid Rain	Degradation of green chlorophyll.
Lead	Neurotoxin (Brain).
Particles	Can cause lung-cancer.
	Waste diseases.

Control Measures

(*i*) Optimum air to fuel ratio (18:1) should be supplied in the combustion chamber of the engine.

(*ii*) Proper servicing of vehicles.

(*iii*) Pollution Check-Up of vehicles (Indian Standards).

(*iv*) Use of High Octane Fuel reduces emission of lead in the atmosphere.

(*v*) Delhi government is providing handsome amount of subsidy on the YO-Bikes which run by electricity chargeable batteries (reducing pollution).

(*vi*) Use of cleaner fuels. *e.g.* CNG, LPG.

10.8 Urbanization and Environment

Industrial cities are set-up as a result of population immigration to urban areas.

☆ As a result of Urbanization, large numbers of industries are set up (industrialization), which release large amount of pollutants to air, water, land (Environmental Pollution) industries also cause noise pollution.

Table 10.2: List of activities which cause environmental pollution

Industrial Activity	*Pollution*
Mining	Air, Water, Land, Noise Pollution.
Modern Agriculture	Air, Water, Land, Noise Pollution.
Construction	Air, Water, Land, Noise Pollution.

Control Measures

☆ **Sustainable Development:** Developmental activities carried out integrating social, economic and environmental factors.

10.9 Population Growth

☆ Rate of growth of population is **exponential** in nature. It puts large amount of pressure on natural resources to fulfill demands of growing population. (House, Food, Clothes and Other Utilities). So the natural resources are getting less in availability.

(Scarcity of Natural Resource) causing the situation Population – Explosion → (Exponential Rise in Population + Scarcity of Natural Resources)

Participatory Learning

(a) Objective Questions

1. The thinning of stratospheric ozone layer during the spring time is called______.
2. Peeling of ozone umbrella, which protects us from UV rays, is caused by______.
3. Formation of ozone hole is maximum over______, DISCOVERED by______.
4. Ozone conservation day OR International Ozone Day is ______(Decided by UNEP: United Nations Environment Programme).
5. The international conference to protect ozone layer is ______.
6. Ozone layer absorbs ______radiation from sun. Ozone layer is present in______. (***UPTU, 2010***)
7. Maximum concentration of ozone found in the layer of atmosphere is______.
8. Ozone is measured in______, 1 DU= 1ppb. Concentration of ozone in Stratosphere is______.(***UPTU, 2010***)
9. CFCs (Chlorofluoro-carbons), CH_4, N_2O, Carbon Tetrachloride, Halon and Methyl bromide are known as______. CFCs are discovered by______.
10. CCl_2F is Freon-11 and CCl_2F_2 is Freon-12. Freon -11 and Freon-12 are the ______of CFCs.
11. Chlorofluorocarbon releases a chemical harmful to ozone called______. (***UPTU, 2009***).
12. Carbon-di-oxide is the______. Carbon-di-oxide contributes ______to the "Green House Effect" on earth.
13. Green House Effect is related to______.(***UPTU 2010***)
14. The radiatively active gases are also called______.

15. The gas with maximum potential of Green House Effect is______.

16. Methane has ______than Carbon-di-oxide.

17. Methane is released from ______fields. Nitrous oxide is also a green house gas provided from agricultural fields.

18. The ______include carbon-di-oxide, methane, nitrous oxide, ozone, water vapour and chlorofluorocarbons.

19. The global warming contributes to rise in ______due to thermal expansion of ocean and melting of ice/glaciers.(***UPTU, 2010***)

20. The green house effect is due to impermeability of ______through CO_2 of the atmosphere. *i.e.* they do not allow long wave radiation emitted from earth to escape in space.

21. The international conference to reduce green house gas emissions is ______.

(b) Very Short Answer Type Questions

1. Discuss the potential of green house gases in global warming.

2. How global climate change is linked to human health.

(c) Short Answer Type Questions

1. Define causes of ozone hole formation.

2. Define green house effect.

(d) Long Answer Type Questions

1. Describe causes, effects, and control measures of ozone layer depletion.

2. Describe causes, effects, and control measures of global warming (***MTU, 2013-14***).

3. Comment upon the global climate change taking place in the environment.

Answers to Objective Questions

1. Ozone hole

2. CFCs

3. Antarctica, Dr. Joe E. Farman

4. 16 Sept

5. Montreal Protocol

6.

   UV, Stratosphere

7. Stratosphere
8. Dobson Units (DU), 8 ppm
9. ODS (Ozone Depleting Substances), Thomas Midgley
10. Trade names
11. Free-radical Chlorine
12. Major Green House Gas, maximum
13. Global Warming
14. Green House Gases
15. Methane
16. 10 times more potential
17. Rice/paddy
18. Green house gases
19. Sea level
20. Long wavelength radiations
21. Kyoto protocol.

Chapter 11

Environmental Education

11.1 Environmental Education

The process of creating awareness among the general public about the local environmental issues is known as **environmental education**. Many environmentalists have focused on environmental conservation. *e.g.* **Sunderlal Bahugna and Chandi Prasad Bhatt** (Chipko Movement), **Anil Agarwal** (journalist, founder of an NGO: Centre for Science and environment, wrote the first report on the 'State of India's Environment' in 1982), **Medha Patkar** (Journalist, Narmada Bachao Andolan), **M C Mehta** (environmental lawyer, cases handled: protecting the Taj Mahal, cleaning up the Ganges River, banning intensive shrimp farming on the coast), **Madhav Gadgil** (ecologist), **Mahatma Gandhi** (father of nation, sustainable lifestyle), **Indira Gandhi** (prime minister of India, environmental conservation based policies and regulations), **S P Godrej** (wildlife conservation and nature awareness programs, the Padma Bhushan in 1999) **Anna Hazare** (peace and anti-crime group, initiated rain water harvesting in India), **Ralph Emerson** (the role of commerce to our environment (1840s)), **Rachel Carson** (published book famous as 'Silent Spring', the adverse effects of pesticides of human health, gained international importance).**Salim Ali's** (environmental conservation and policy-making). Justice "**Kuldeep Singh**" known as green judge declared environmental sciences as mandatory subject in all graduate courses.

11.2 Principles of Environmental Education (*MTU, Carryover, 2012-13*)

Intergovernmental conference on environmental education, Tbilisi, USSR, 1977 described the objectives of environmental education as awareness, knowledge (understanding); attitudes, skills (appreciation); participation (active public participation).

Process of creating Environmental Education

The sequential steps of environmental education are given as follows (**Figure 63**):

i) Understanding

First, the public will understand the importance of local environmental issues. *e.g.* importance of clean surroundings and clean water-bodes.

ii) Appreciation

The public will start appreciating, the importance of environmental conservation.

iii) Active Public Participation

After understanding and appreciating, the public will be actively involved in environmental conservation. *e.g.*, afforestation (Table 11.1).

Process of Environmental Education

Awareness of local surroundings → Public will understand the importance of its conservation

⇩

Appreciation

⇩

Active involvement of public in environmental conservation
e.g. Chipko movement

Figure 11.1: Sequential Steps of the Process of Environmental Education.

11.3 Need for Public Awareness or Importance of Environmental Education: Active public participation

The local public get actively involved in environmental conservation, after understanding the importance of environmental conservation *e.g.* waste management, population control, afforestation (planting trees), encouraging renewable energy sources, rural development, saving energy resources in our daily life.

11.3.1 Details of Environmental Movements Indicating the Importance of Environmental Awareness

11.3.1.1 Chipko Movement in Tehri Garhwal Gained International Importance

Bishnoi People hugged the trees to stop cutting of trees (deforestation). The movement was led by Sunderlal Bahuguna and Chandi Prasad Bhatt. The movement was recognized at international level. This showed the importance of public awareness (**Figure 11.2**).

Company started cutting trees → in Tehri Garhwal | Local people started fighting against cutting of trees

They hugged the trees

National Leader: Sunderlal Bhaguna

Figure 11.2: Chipko Movement.

Table 11.1: List of environmental awareness movements

Environmental Movement	*Description*	*Outcome : Environmental Conservation*
Bishnoi movement in Rajasthan(*UPTU, 2009*): 20 (Bis; twenty) and 09 (Noi, nine) principles in the community of guru Jambheshwar	Local people of the Bishnoi community protested against cutting of Khejari trees by the order of the then king of Jodhpur	**"Role of women in environmental conservation" Amrita Devi,** local woman, played a key role, gained publicity.
Silent valley movement in Kerela	Construction of a dam on the river Kuntipujha, in the biodiversity rich forest area of silent valley. An NGO "Kerela Sastra Sahitya Prishad (KSSP)" protested against the dam construction.	Government declared it a national park in 1984.
Beej Bachao Andolan	Vijay Jardhari and local people of Henwal River Valley in Tehri district (Uttarakhand, India), in late 1980's initiated a movement to protest against HYV seeds and saving traditional variety to seeds to ensure self-sufficiency of farmers in agriculture and safe production to fulfill the requirements of all the individuals of present generation and saving the quality of air, water and soil for nurturing future generations known as sustainable agriculture.	(i) **Conserving forest resources**. (ii) **Conserving biodiversity:** Traditional variety of seeds contains various desired characters *e.g. drought resistance, pest attack tolerance, ability to grow at high temperature, wholesome nutrition and taste.* (iii) **Minimizing ill-effects of modern agriculture**.

11.4 The 'Guidelines for Excellence' for Environmental Education (*UPTU, 2011*)

North American Association for Environmental Education (NAAEE) has described guidelines for excellence in environmental education 1996 (revised 2004). Environmental education material should follow following important points:

(i) Accuracy

Material of environmental education should be accurate in describing environmental problems.

Precise (right/accurate) environmental education material (newspaper, magazine, nukkad natak, pamplets *etc.*)	→	**local environmental problems**

(ii) Depth

Understanding of environmental concepts, conditions and issues, awareness of feelings, values and attitudes of environmental sciences explains the depth of the knowledge.

(iii) Skills Building

Critical thinking, sound communication help build a feeling of appreciation among the people.

(iv) Action Orientation

Active involvement of people is necessary to generate active public participation towards environmental conservation.

(v) Soundness of Instructions

Instructions should be learner friendly so that effective learning environment is created.

(vi) Usability

Environmental education material should be easy to use.

11.5 Process of Public Awareness

Understanding, appreciation and active public participation is the sequence of basic methodology of the process of creating awareness among local people about local environmental issues.

Methods or Means of Environmental Education

(i) **Media:** Television, radio, newspapers, pamplets, magazines (Down to Earth by Centre for Science and Environment)

(ii) Nukkad-Natak in villages

(iii) Websites

(iv) Ecoclub programmes in schools.

(v) Organizing various competitions at national and international levels with the theme on environment *e.g.* drawing & painting competitions, poem recitation, essay, speech.

(vi) Conferences, workshops, seminars.

11.6 Agencies Active in Creating Public Awareness (Environmental Education): Non-Governmental Organizations

NGO are the non-profit making organizations working towards the protection of environment and upliftment of society (**Table 11.2**). They help create awareness among the local people by various following methods:

Table 11.2: Examples of NGOs.

NGO	*Location*	*Function/Role*
Centre for Science Env. (CSE)	Delhi (Director: Ms. Sunita Narain)	Environmental awareness by research and field work.
Centre for Environment Education (CEE)	Ahmedabad (Director: Dr. Vikram Sarabhai)(All India Branches and International Offices)	Environmental awareness at local level Eco-clubs in school. Green Corps program. Joint Forest Management (JFM)
World wide fund of nature (WWF)	New Delhi (Symbol: Giant Panda, an extinct animal)	Awareness for conservation of wild animals.
Participatory Research in Asia (PRIA)	Delhi	Secondary research
Kalpavriskh	Pune (Maharashtra)	Narmada Bachao Andolan Its activities include talks and audio-visuals in schools and colleges, nature walks and outstation camps, organising student participation in ongoing campaigns including street demonstrations, pushing for consumer awareness regarding organic food, press statements, handling green alerts,and meetings with the city's administrators. It is involved with the preparation of site-specific, environmental manuals for school teachers. Kalpavriksh was responsible for developing India's National Biodiversity Strategy and Action Plan in 2003
Greenpeace	Amsterdam, and 28 national and regional offices in the world, (showing a presence in over 40 countries). It is a **global environmental organisation**.	✰ Efforts to conserve marine ecosystem, from oil pollution ✰ Emphasized 'green budget' or ecologically sustainable budget 2011-2012 :(i) to promote renewable energy sources. (ii) to promote ecological agriculture.(iii) An online signature campaign started to hold public consultation before decide for mining in forests. 50,000 people supported saving forests and not continuing mining.
CPR Environmental Education Centre, Madras		Promote conservationof nature and natural resources among NGOs, teachers, women, youth and children to generally

Contd...

Table 11.2–***Contd...***

NGO	*Location*	*Function/Role*
Salim Ali Center for Ornithology and Natural History (SACON), Coimbatore	Dr. Salim Ali's dream that became a reality only after his demise. He wished to support a group of committed conservation scientists	Has instituted a variety of field programs that have added to the country's information on our threatened biodiversity.
Uttarkhand Seva Nidhi (UKSN), Almora		Sustainable resource use at the village level through training school children. Its environment education program covers about 500 schools

- ☆ Media.
- ☆ Trainings and Workshops.
- ☆ Nukkad Natak.
- ☆ Magazines.
- ☆ Campaigns.
- ☆ Awareness Programs (Posters, Slogans, Competitions etc).

NGOs Work as Catalyst

NGO are the Non-Governmental Organizations creating awareness among public to save their local environment. It helps the local people by providing economic benefits, knowledge by magazines, newspapers, television *etc.*, livelihood (job) benefits and incentives.

Finally motivate them for active public participation to save their local environment. They do not take part in the process of environmental conservation. *e.g.* **Joint Forest Management (JFM)**

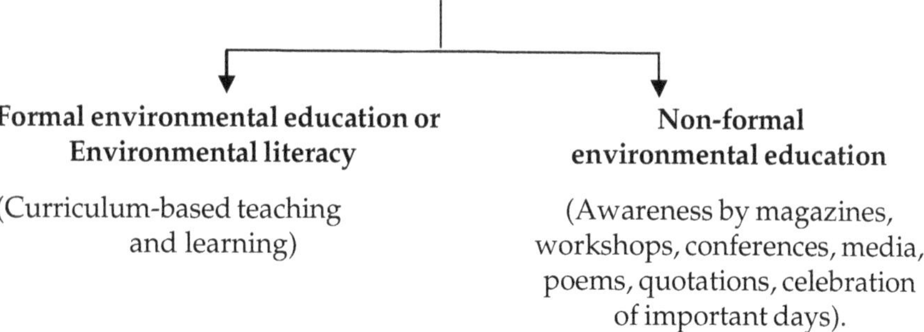

11.7 Conferences, Workshops, Seminars

Environmental Calender: Important Environmental Conservation Dates

(Basic agenda is environmental education)

☆ National Science Day	28 February
☆ World forestry Day	21 March
☆ World Resource Day	10 April
☆ World Earth Day	22 April (***UPTU, 2007***)
☆ International Energy Day	04 May
☆ World Environment Day	5 June (***UPTU, 2006***)
☆ World Population Day	11 July
☆ India's Independence Day	15 Aug, 1947
☆ World Ozone Conservation Day	16 September
☆ Engineer's Day	15 September

Table 11.3: Environmental Education by Environmental Conferences.

Environmental Conferences	*Important features*
The Stockholm Conference 1972 at Stockholm	☆ Ozone layer depletion ☆ 5 June as World Environment Day ☆ Integrated environmental, economic and social development.
Tbilisi, USSR 1977	☆ Intergovernmental conference on environmental education ☆ Awareness, knowledge (understanding); attitudes, skills (appreciation); participation (Active Public Participation)
Rio declaration on Environment and development (The Earth Summit) 1992	☆ Economic progress is directly linked with environmental protection. ☆ Equitable global partnership involving governments, populations and key sectors of societies is emphasized. ☆ Developing international agreements that protect the integrity of the global environment and the development system. ☆ Major international treaties and agreements emphasized are global climate change, biological diversity, deforestation, and desertification.
The Biodiversity Convention	☆ Conservation and sustainable use of biological diversity ☆ (i) *In-situ* conservation (ii) *ex-situ* conservation ☆ (iii) Potential risks associated with the handling and introduction into the environment of living modified organisms (LMOs) are addressed. The need to promote bio-safety issues is defined.
United Nations Framework Convention on Climate Change (Kyoto Protocol. Dec. 11, 1997)	Different goals and commitments to reduce emission of greenhouse gases.
World Summit on Sustainable Development, (WSSD, 2002) at Johannesberg, South Africa	☆ "Think globally Act Locally" ☆ Poverty alleviation (social, economic and environmental equity)
Ramsar Convention (The Convention on Wetlands, Iran, 1971)	☆ Conservation of wetlands ☆ Intergovernmental treaty: local, regional and national actions and international cooperation, as a contribution towards achieving sustainable development.
Rio +20 Earth Summit (June 2012 at Rio De Janerio)	☆ Discussed issues of sustainable development. ☆ No concrete outcome. ☆ Concern for environment.

- ☆ World Habitat Day — 03 October
- ☆ World Wildlife day — 06 October
- ☆ World Food Day — 16 October
- ☆ World Environment Protection Day — 26 November
- ☆ World AIDS Day — 1 December (***UPTU, 2008***)

Participatory Learning

(a) Objective Questions

1. Environmental education should be imparted only at

   (a) Primary school stage (b) Secondary school stage

   (c) College stage (d) All stage

2. World environment day is ______.

3. World ozone day is ______.

4. World AIDS day is ______. (*UPTU, 2008*).

5. Chipko movement was initiated in Tehri Gahrwal to protest against ______ by ______ them.

(b) Very Short Answer Type Questions

1. What is the agenda for chipko movement?

2. What is the role of environmental education in environment protection? Or importance of environmental education.

3. Define Non-Governmental Organizations.

4. Define eco-mark or eco-labeling. What is the eco-mark of India called?

5. Define eco-clubs.

(c) Short Answer Type Questions

1. We celebrate ozone day, earth day, environment day *etc*. What is the basic agenda?

2. Describe the role of environmental movements in environmental conservation.

3. What are the basic principles of environmental education? ***(MTU Carryover, 2013)***.

4. Comment "NGO acts as catalyst".

(d) Long Answer Type Questions

1. What are the guiding principles of excellence in environmental education (***MTU, 2010***).

2. Describe the role of NGOs in environmental protection (***MTU, 2012***).

3. Describe the salient features of Forest Conservation Act, 1980 and explain its amendments (***MTU Carryover, 2013***).

4. Give an account of type of environmental education in our country, India.
5. What are the goals of environmental education programme launched by UNESCO in 1975?
6. Describe the need or importance of public awareness.
7. Does environmental education is a tool to sustainability? How?

Answers to Objective Questions

1. (d)
2. 5 June
3. 16 September
4. 1 December
5. Cutting of trees, hugging

Chapter 12

Social Environmental Issues

12.1 Social Environmental Issues are Important for Sustainable Developmental Strategies

Solving the social environmental issues help in strengthening sustainable developmental framework.

12.1.1 Watershed Management

Water crisis is the major problem of agriculture sector due to limited scope of expansion of irrigation potential, non-judicious use of existing water resources, continuously declining potential of reservoirs, poor efficiency of irrigation projects, greater conveyance losses, diversion of water from agriculture to non-agriculture purposes and increasing ground water pollution. Proper watershed management can check not only further degradation of ecosystems but degraded land can also be restored (**Figure 12.1**).

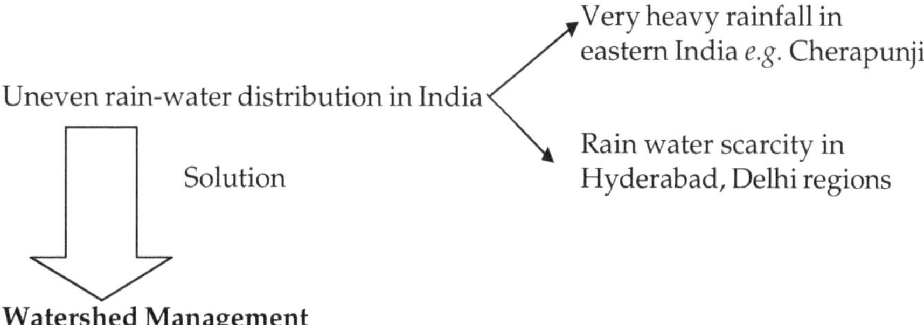

Watershed Management

(Efficient management of air, water and land resources to achieve sustainable water-supply system)

Figure 12.1: Need for Rainwater Harvesting in India.

12.1.1.1 Rainwater Harvesting

Traditional rainwater harvesting techniques through community participation entails wider perspective. Traditionally rainwater is stored in tankers built in the countryyard and artificial wells called kundli. Monsoon runoff was captured in zings in Ladhakh, achars in Bihar, johads in Rajasthan and Eris in Tamil Nadu. Drought accelerates the rate of ecosystem degradation. Rangelands were being used for grazing three times more than their peak capacities in a non-drought year, resulting in severe degradation as well as accelerated soil erosion. Sustainable agricultural techniques not only reduce the pressure on land resources but also on water resources resulting in sustainable yield of food (**Figure 12.2**).

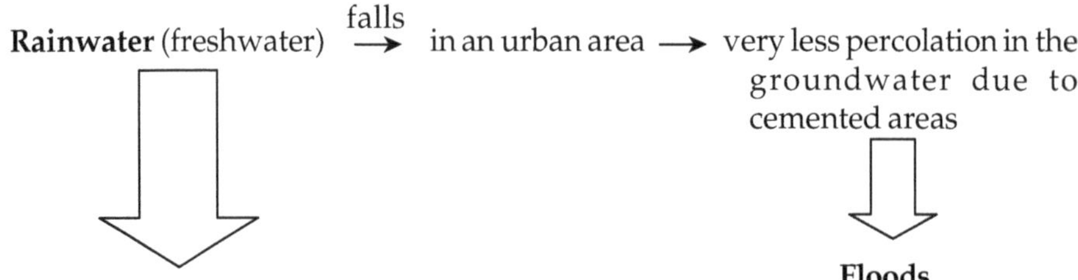

Watershed Management
Concept: Capture the rainwater where it falls.
(*Do not let single rain-drop waste*)

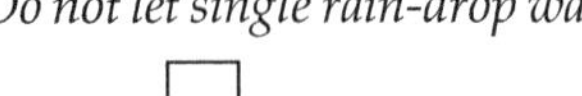

Tool of watershed management: **Rainwater Harvesting**
At the roof-top of houses in urban areas, rainwater is stored.
Anna Hazare ji is the national leader to promote rainwater harvesting.
Traditional Rainwater Harvesting: Baolis, Kunds

Figure 12.2: Importance of rainwater harvesting

12.2 Flyash Landfill Reclamation

Planting tolerant trees on the flyash landfill dykes is the most viable environmental option *e.g.* Arjun, neem, people, bargad, *Bauhinia variegata* are the tolerant tree species which can absorb large amount of heavy metal and detoxify them.

12.3 Disaster Management

The processes of forecasting, predicting and assessment, analyzing, prevention and mitigation and relief and rehabilitation measures are the important steps of disaster management. Important disasters of concern are drought, flood, volcanoes, cyclones and earthquakes (**Figure 12.3**).

12.3.1 Urban Disasters

The vulnerability of mankind to disasters of various types has increased considerably all over the world. India is vulnerable to varying degree of natural and

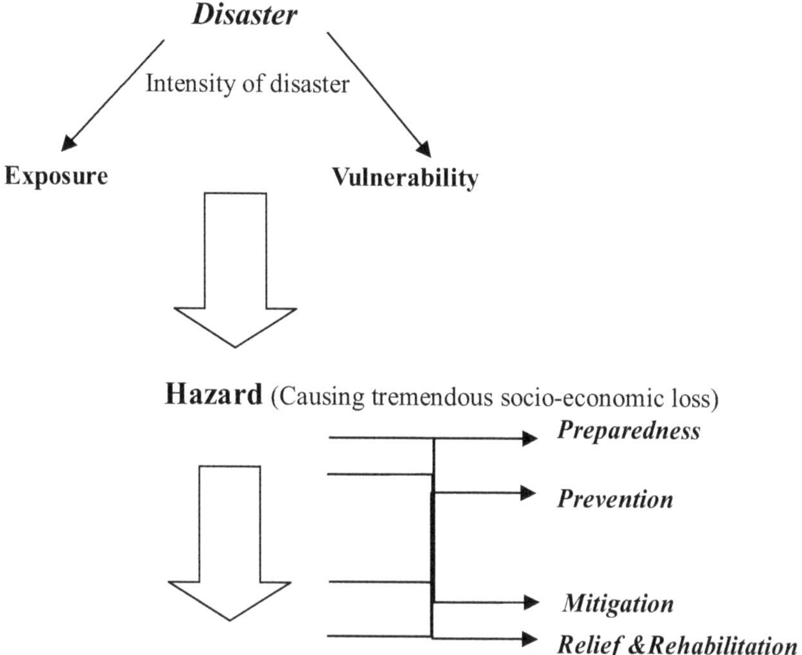

Figure 12.3: Need for Disaster Management.

man-made disasters. About 55 per cent of India's landmass is prone to earthquakes of varying intensity, 68 per cent is vulnerable to drought, 12 per cent to floods, 8 per cent to cyclones apart from heat waves and severe storms. Disasters are natural catastrophes as well as Man-made emergencies.

India has diverse geo-climatic conditions which make the country among one of the most vulnerable to natural disasters in the world. Disasters occur with very high frequency in India and while the society at large has adapted itself to these regular occurrences, the economic and social costs continue to mount year after year. It is highly vulnerable to floods, drought, cyclones, earthquakes, landslides *etc*. Moreover the pace of these disasters has been induced by various anthropogenic activities including climate change. Almost all parts of India experience one or more of these events. An attempt has been made in this section to outline some of the disasters and various laws that deal with pre, during and post disaster management.

12.3.1.1 Flood

India is highly vulnerable to floods and out of total geographical area of 329mha, more than 40 mha are flood prone. Floods, which are recurrent phenomenon, cause huge loss and damage to livelihood system, property, infrastructure and other public utilities. The average annual flood damage during 1996-2005 was estimated to be Rs.4745 crore as compared to Rs.1805 crore the corresponding average for the last 53 years (NDMA guidelines for Flood Management). This figure clearly shows increasing trend in the occurrence of floods which has been accelerated due to population growth, industrialisation, urbanisation, deforestation and other activities creating mismatch between environment and economic growth. Floods being the most common natural disaster, people have out of the experienced devised many coping mechanisms but more considerable efforts are required and put in place a techno-legal regime to make structures flood proof and regulate the activities in the flood plains of the rivers.

12.3.1.2 Cyclone

India has a coastline of 7,516 km, of which 5,700 km are prone to cyclones of various degrees. About eight per cent of the Country's area and one-third of its population live in 13 coastal states and UTs who are, thus vulnerable to cyclone related disasters. Loss of lives, livelihood opportunities, damage to public and private property and severe damage to infrastructure are the resultant consequences, which can disrupt the process of development. Climate change and the resultant sea-level rise are also likely to exacerbate the seriousness of this problem in the coming decades. The National Guidelines on Cyclone Management, 2008 states that since 1737,21 of the 23 major cyclone disasters (in terms of loss of lives) in the world have occurred over the Indian subcontinent (India and Bangladesh). Tropical cyclones (wind velocity > 61 km/h) in the Bay of Bengal striking the east coast of India and Bangladesh usually produce a higher storm surge as compared to elsewhere in the world because of the special nature of the coastline, shallow coastal bathymetry and characteristics of tides. Their coastal impact is significant because of the low flat coastal terrain, high density of population, low awareness of the community, inadequate response and preparedness and absence of any hedging mechanism.

12.3.1.3 Landslides

Our country experiences landslides year after year especially during the monsoons and periods of intense rain. This hazard affects about 15 per cent of our country covering over 0.49 million square kilometres. Landslides of different types occur frequently in the geodynamically active domains of the Himalayan and Arakan-Yoma regions, as well as in the relatively stable domains in the Meghalaya Plateau, the Western Ghats and the Nilgiri Hills. Besides, sporadic occurrences of landslides have been reported in the Eastern Ghats, Ranchi Plateau, and Vindhyan Plateau as well. In all, 22 States and parts of the Union Territory of Pudducherry and Andaman & Nicobar Islands of our country are affected by this hazard, mostly during the monsoons. Extensive anthropogenic interference is a significant factor that increases this hazard manifold. Landslide disasters have both short-term and long-term impact on society and the environment. The short-term impact accounts for loss of life and property at the site and the long-term impact includes changes in the landscape that can be permanent, including the loss of cultivable land and the environmental impact in terms of erosion and soil loss, population shift and relocation of populations and establishments.

12.3.1.4 Earthquakes

India's high earthquake risk and vulnerability is evident from the fact that about 59 per cent of India's land area could face moderate to severe earthquakes. During the period 1990 to 2006, more than 23,000 lives were lost due to 6 major earthquakes in India, which also caused enormous damage to property and public infrastructure. in the 1990s, India witnessed several earthquakes like the Uttarkashi earthquake of 1991, the Latur earthquake of 1993, the Jabalpur earthquake of 1997, and the Chamoli earthquake of 1999. These were followed by the Bhuj earthquake of 26 January 2001 and the Jammu & Kashmir earthquake of 8 October 2005. In most earthquakes, the collapse of structures like houses, schools, hospitals and public buildings results in the widespread loss of lives and damage. Earthquakes also destroy public infrastructure like roads, dams and bridges, as well as public utilities like power and water supply installations. Past earthquakes show that over 95 per cent of the lives lost were due to the collapse of buildings that were not earthquake-resistant. Though there are building codes and other regulations which make it mandatory that all structures in earthquake-prone areas in the country must be built in accordance with earthquake-resistant construction techniques, new constructions often overlook strict compliance to such regulations and building codes.

12.3.1.5 Drought

Around 68 per cent of the geographical area of the country is prone to drought in varying degrees. In India, drought occurs mainly due to the failure of South-West monsoon (from June to September). There is lot of variation of rainfall both in terms of area, extent of occurrence, precipitation, intensity and time of occurrence. More than 73 per cent of annual rainfall is received during the South West Monsoon. Hence, areas, which have received less rainfall during this period and affected by drought needs to wait till the next monsoon. Droughts in India are always linked with the performance of the monsoon. In the course of the last 125 years, Indian had at least 38 years of poor rainfall. In more recent times, there were four major droughts periods – 1965-66, 1972, 1987 and 2002, although droughts in one part or the other of this vast sub-continent is almost an annual feature. Drought is a natural hazard that differs from other hazards since it has a slow onset, evolves over months or even years, affects a large spatial extent, and cause little structural damage. Its onset and end and severity are often difficult to determine. Like other hazards, the impacts of drought span economic, environmental and social sectors and can be reduced through mitigation and preparedness.

12.3.1.6 Hazardous Chemicals

The growth of chemical industries has led to an increase in the risk of occurrence of incidents associated with hazardous chemicals (HAZCHEM). Common causes for chemical accidents are deficiencies in safety management systems and human errors, or they may occur as a consequence of natural calamities or sabotage activities. Chemical accidents result in fire, explosion and/or toxic release. The nature of chemical agents and their concentration during exposure ultimately decides the toxicity and damaging effects on living organisms in the form of symptoms and signs like irreversible pain, suffering, and death. Meteorological conditions such as wind speed, wind direction, height of inversion layer, stability class, *etc.*, also play an important role by affecting the dispersion pattern of toxic gas clouds. The Bhopal Gas tragedy of 1984—the worst chemical disaster in history, where over 2000 people died due to the accidental release of the toxic gas Methyl Isocyanate, is still fresh in our memories. Such accidents are significant in terms of injuries, pain, suffering, loss of lives, damage to property and environment. Increased industrial activities and the risks associated with HAZCHEM and enhanced vulnerability lead to industrial and chemical accidents. Chemical accidents may originate in the manufacturing or formulation facility, or during the process operations at any stage of the product cycle, material handling, transportation and storage of HAZCHEM. There has been a paradigm shift in the government's focus from its rescue, relief, andrestoration-centric approach to a planning, prevention/mitigation and preparedness approach.

Indian disaster management policy, 2009 is geared to make a paradigm change from response and calamity relief to disaster prevention, preparation and mitigation. Another significant change is to move from disaster management largely from government to public private partnership, and community disaster management. The efforts have been made to move from disaster management to disaster risk management and finally disaster risk reduction.

12.4 Animal Husbandry

The branch of science which deals with domestic animal rearing and breeding under human care, is known as animal husbandry.

Important Aspects of Animal Husbandry

- Animal husbandry is also known as Animal Science, Stock Breeding, Dairy Farming or Husbandry.
- It plays important role in national economy and in the socio-economic development of the country.

Advantages

(i) Animals get human care.

(ii) Saving animal from getting extinct.

(iii) We get useful products from animals.

Animals	*Products*
Cow	Milk
Sheep	Wool
Goat	Milk
Poultry	Egg

Diadvantages

The ability of animals to adapt to the natural environment decreases.

12.5 Women Education

Examples

- Medha Pathekar- famous environmental journalist.

- ☆ Amrita Devi- started Bishnoi movement in Bishnoi Village.
- ☆ Indira Gandhi: Former Prime minister worked for sustainable development of India
- ☆ Lalita Ramdas was one of the "peace women", brought Greenpeace to India.

Importance of Women in Sustainable Development

- ☆ Women are the half of the human population.
- ☆ Women education is most important part of Sustainable Development.

Problems in Women Education

(i) Under nourishment: Less food is available.
(ii) Mal nourishment: Lack of nutrients in food.
(iii) Sexual harassment.
(iv) Lower status in society.
(v) Forced for getting more child.
(vi) Early marriages in rural sector.

Table 12.1: Policy, Programmes and Strategies for Women Welfare.

Policies or Programs	*Women Welfare*
Sarwa Siksha Abhiyan	☆ National mission
	☆ Quality education for 8 years (6 to 14 years of age).
Balika Samridhi Yojana	
Indira Mahila Yojana	☆ Providing small jobs or income
	☆ Self-dependency
Mahila Samridhi Yojana	
Ladly Yojana	☆ By Delhi Govt.
	☆ 1 Lakh Rs to the poor-parents (income is below poverty-line) of new-born girl child
	☆ UNESCO recommends women education strongly.
Programme of development of Women	To improve socio-economic status of women in rural areas.

12.6 Environmental Problems in India (Developing Countries)

Uttar Pradesh most populated state of country, covers 236286 square kms. Lucknow is designated as its capital. Ganga, Yamuna, Ramganga, Gomti and Ghaghra are the major rivers. As per the census of 2001, a population of 166052859 resides in the state. Female and male populations ratio rounds to 898:1000. Population density is calculated as 689 persons per square km.

(*i*) Natural Resources

- ☆ Poverty-magnifies hunger and malnutrition. Inequitable access to available food further compounds the problem in U.P. Natural resources are exploited. *Viz.* Non Timber forest products, fuel wood. 40.85 per cent people are living below poverty line. (Ref. Economic Survey of India). Electrified villages count to 77047 out of 112803 total number of villages.
- ☆ Disasters frequently occur in the state. Common are floods and droughts. This also indicates the need for integrated water resources management systems.
- ☆ Need to emphasize on Overgrazing, hunting, unsustainable aquaculture.
- ☆ Biodiversity (Flora and Fauna) conservation needs more careful and attentive steps.

(*ii*) Water

- ☆ The integrated water distribution management system is a thrust issue. Percentage of household having access to drinking water is 85.6.
- ☆ Water Management Systems are to be strengthened.
- ☆ All mismanagement leads to Water pollution.

(*iii*) Illiteracy

- ☆ Literacy rate is 57.36 per cent as per census in 2001.Due to lack of knowledge, people consider more family members as an asset to generate more income. This leads to population increase, Unemployment.
- ☆ For the lure of Son, women are compelled to give births to more wards.

(*iv*) Agriculture

- ☆ **Unsustainable agriculture practices**

  In rural villages farmers are not aware of exact economically viable systematic and environmental friendly tools for agriculture. For example overuse of pesticides, fertilizers, improper irrigating tools.
- ☆ U.P. has a vast area of land under sodic soil, which is very unfertile.

(*v*) Health and Hygiene

- ☆ Waste disposal is a thrust problem. People are not very conscious regarding segregation of different types of wastes. *viz.* Biodegradable and Non-biodegradable; Toxic hazardous wastes generated in homes (used batteries, paints, used cans of pesticides etc.) Shopkeepers do not bother to place proper dustbins in front of their shops. Even the sweepers do not bother a lot while collecting wastes. Health related problems are thus magnified.
- ☆ 21.2 per cent children were vaccinated during 1998-1999. (Ref. National family health survey)

(*vi*) Environmental Pollution

☆ Traffic norms are not properly followed. Roads are very congested and less maintained. Vehicles are not well maintained, releasing noxious gases and unwanted noises out of their wear and tear.

☆ Industrial pollution (air, water, land and noise pollution) is posing threat to nearby people residing in the area.

☆ **Garbage Disposal**

There are no stringent laws regarding the household waste thrown at improper sites by general public. Wastes spread in open attract insects; cause unpleasant odors, pathogenic infection spread thereby causing diseases.

☆ **Automotive Pollution**

There is no system to look after the check-up of the vehicles. Only the issue of a certificate does not specify the vehicle in proper working condition with emissions under the pollution control norms.

☆ **Decentralization of Enforcement Machinery**

Authority to take action needs to be decentralized, common and easily available to people. Direct Involvement of public in decision making is highly required.

☆ **Noise Pollution**

Various activities cause spread of irritating and harmful noise vibrations in the environment. Viz: Use of loud speakers, General noise caused, Listening music at a louder pitch in residential and sensitive areas. People also blow horns while driving at a very high pitch unnecessarily most of the times.

☆ **River Restoration**

Ganga action plan is executed to regain the Ganga river water quality, degraded due to effluent discharge, and throwing of variety of wastes in the river.

(*vii*) Natural Disasters

☆ Occurrence of floods, landslides is regular feature in India which emphasize the need of effective disaster management.

Participatory Learning

(b) Objective Questions

1. Conserving rain-water is an important technique of ______.

2. Balika samridhi yojana was to impart ______.

`aster management includes ______, ______, ______ and ______ of `rs.

(b) Very Short Answer Type Questions

1. What is the agenda for women welfare?
2. What is the role of government in women welfare?
3. Define human values and human-rights.

(c) Short Answer Type Questions

1. Describe the role of women education in sustainability.
2. Describe the importance of rain water harvesting in water conservation.
3. Describe the techniques of rain water harvesting.

(d) Long Answer Type Questions

1. How social issues are important in sustainable development.
2. Describe the policies for women welfare.

Answers to Objective Questions

1. Watershed management
2. Women welfare
3. Forecasting, predicting, mitigation and relief and rehabilitation.

Chapter 13

Human Population

13.1 Human Population

The term population refers to the group of individuals of a species (similar group of individuals which can interbreed freely among themselves) living a definite geographical area at a given time.

13.2 Population Growth Rate

Population growth rate is measured in terms of annual average growth rate (in percentage).

$$\text{Average annual growth rate} = \left(\frac{P_2 - P_2}{P_1 \times N}\right) \times 100$$

P_1 is the population size measured in the previous census.

P_2 is the population size measured in the present census.

N is the number of years between the two census of population size measurement.

13.2.1 Factors Affecting the Growth Rate

There are certain deciding factors for the growth measurement:

(i) **Birth rate or Natality**: Number of infants produced per thousand individuals depends upon the fertility rate of the population. Average number of children born to a woman in her lifetime, considering age specific birth rate of a given year is known as **Total fertility rate**.

(ii) **Death rate or Mortality**: Death of number of individuals per thousand people. Health care has reduced the death rate causing increase in population growth rate.

(iii) **Migration**: Movement of individuals from one country to other or one region to other region in a country is known as migration.

13.2.2 Population Explosion

Rate of increase of human population is exponential in nature and puts tremendous pressure on natural resources to fulfill the growing demand of natural resources known as natural resource crisis and population explosion.

13.2.3 Causes of Population Explosion

Factors which cause natural resource crisis and population increase:

(i) **Urbanization**: production of variety of things puts pressure on natural resources described in chapter 2.

(ii) **Deforestation**: Most of the urbanization requires clear land. Forest land is cleared for non-forest purposes.

(iii) **Health care facilities**: Increased medical facilities helped human population to recover from many hazardous diseases *e.g.* dengue, cholera, polio, hepatitis.

13.2.4 Effects of Population Explosion

(i) **Environmental pollution**: Increasing industrialization due to population increase causes environmental pollution.

(ii) **Natural resource crisis**: Population increase causes ground water depletion, power crisis, biodiversity loss, deforestation and other problems of natural resource crisis.

(iii) **World food problems**: Undernourishment, malnourishment and over nourishment are the world food problems.

(iv) **Unemployment**: Increasing population would suffer from less availability of jobs.

(v) **Poverty**: Increasing population leads to unequal distribution of natural resources leady to poverty.

13.2.5 Control Measures of Population Explosion

(i) **Education**: Education plays very important role in control of population in women.

(ii) **Environmental awareness**: Creating awareness among general public about the importance and measures of birth control is an important step.

(iii) **Government policies**: Government provides incentive schemes only upto two children in a family *e.g.* reservation in higher education in top institutes (IITs, medical colleges, DCE) for the single girl child to promote birth control among the parents having first daughter. Maternity leaves facilities are available only upto giving birth to two babies.

(iv) **Economic welfare**: Economically better societies have population with fewer children.

13.3 National Population Policy, 2000 (NPP, 2000)

It aims at control of population by providing the needs of contraception, health care and achieving a stable population by 2045 to ensure sustainable development.

13.4 Environment and Human Health (*UPTU, 2007, 2009*)

Population control is the best method to ensure good human health so that the medical facilities may be made available to total population, natural resources can be sustainably utilized and causing less environmental pollution.

13.4.1 Sexually Transmitted Diseases (STD)

AIDS, hepatitis, syphilis, gonorrhea, chlamydiosis and cervical cancer are the sexually transmitted diseases, more commonly found in poor countries like Africa and among workers.

13.4.1.1 HIV/AIDS

Acquired Immuno Deficiency Syndrome (AIDS) is caused by Human Immuno deficiency Virus (HIV) Blood transfusion and organ transplantation, unprotected sexual contact, using contaminated needles of injections and breast milk feeding are the causes of transmission of disease from the infected person to a healthy person. Swoolen lymph nodes, regular fever, sweating at night, weight loss, severe brain damage leading to memory loss, loss of ability to speak and clearly think and decreased platelet counts are the symptoms of AIDS. Anti-HIV antibodies in the blood of the patients are detected by ELISA test (Enzyme linked immune-sorbent assay). Western blot test confirms a positive ELISA test. Sustainable lifestyle is an important cure of AIDS by taking care of the causes.

13.5 Role of IT in Environment and Human Health

Remote sensing and geographic information system are the important tools of IT which help in environmental conservation.*e.g.* in gathering information of biodiversity and prediction and forecasting of disasters (like volcanoes, earthquakes, floods, disasters). Agro-advisory boards are established which predict the weather scenario as per the requirement of agriculture.

13.6 Human Rights

These are the fundamental rights of human beings to live peacefully in the biosphere, also known as ecological justice. United Nations Organisation has declared Universal Declaration of Human Rights (UDHR) which provides comprehensive protection to all the individuals to life, protection, security, freedom of thought, expression, fair trial by law and freedom of association and movement.

13.7 Value Education

Ethical values are implicit in the education framework including compassion, helpfulness, peace loving, generosity, adaptability, politeness and tolerance. These ethical values are also called as environmental ethics and forms the basics of sustainability. The principles of sustainability: intergenerational equity, intergenerational equity and ecological justice are important points which emphasize conserving nature and living organisms is very important.

Participatory Learning

(a) Objective Questions

1. Population increase causes ______.
2. Death rate is measured as______
3. Birth rate is measured as ______.

(b) Very Short Answer Type Questions

1. What is population explosion?
2. What is the role of population in growth and development, comment both ways?
3. Define values education.

(c) Short Answer Type Questions

1. Describe the role of 3 Ps in environmental degradation. (***MTU, 2013-14***).

(d) Long Answer Type Questions

1. Explain the causes, effects and control measures of population explosion.

Answers to Objective Questions

1. Natural resource crisis and environmental pollution
2. Mortality
3. Natality

Previous Papers

B. Tech (Sem I) Theory Examination, 2012-13, Environment and Ecology Paper Code: CE 101

Time: 3 Hours] *[Total Marks: 80*

Note: Attempt **all** questions from all sections as per instructions.

Section A

Attempt all parts of this question. **2x8=16**

1. (a) What are the components of the earth or environmental segments?

(b) Comment on human concern on environment.

(c) What are the impacts of *in-situ* and *ex-situ* conservation of biodiversity?

(d) Trace the factors leading to deforestation.

(e) Classify the energy resources.

(f) What is the composition of Biogas?

(g) What are the symptoms of water pollution in any water body?

(h) List the effects of air pollution on public health.

Section B

2. Attempt any **three** parts of this question. **8x3=24**

(a) Explain the following:

(i) Ecosystem and its components

(ii) Functions of ecosystem

(b) What is sustainability and sustainable development? What are its components? How can it be ensured?

(c) Depletion of power production is the major problem faced in our country. Provide reasons for this problem and also give your innovative ideas on alternate source of energy for electricity production.

(d) How the water gets polluted? Explain each pollutant with example.

(e) Explain in detail the various acts relating to air, water and environmental protection.

Section C

Note: Attempt all questions of this section: **8x5=40**

3. Attempt any two parts:

(a) Prove why conservation of forest is important to mankind.

(b) Compare natural and man-made extinctions with examples.

(c) Why sustainable agriculture is needed?

4. Attempt any one part: **8x1=8**

(a) Elaborate environment and its importance by showing few time examples?

(b) Natural cycles play an important role in our environment. Explain any two of them.

5. Attempt any one part: **8x1=8**

(a) What are the products of combustion of coal and petroleum? Discuss their damaging effect on environment.

(b) Give critical notes on the following:

(i) Tidal wave energy

(ii) Geothermal energy.

6. Attempt any one part: **8x1=8**

(a) What is meant by " water shed"? Explain with reasons the importance of watershed management in our national policy.

(b) "Day by day atmosphere gets prone to serious effects" Prove this statement.

7. Attempt any two parts: **4x2=8**

(a) Discuss the Bhopal gas tragedy.

(b) What are the various water-borne diseases.

(c) Illustrate the activities of NGOs on environmental protection.

B. Tech (Sem II) Theory Examination, 2011-12, Environment and Ecology Paper ID 3103

Time: 2 Hours] *[Total Marks: 50*

Note: Attempt **all** questions from all sections as per instructions.

Section A

1. Attempt all questions which are compulsory. **5x2=10**

(a) What is "Environment"/How can you say that it constitutes as " life supporting system"?

(b) Enumerate the services provided by a healthy ecosystem.

(c) What do you understand by a "Natural Resource"? What are its kinds?

(d) Give a brieg account of non-fuel mineral resources in India.

(e) What are the conventional sources of energy for the humanity?

Section B

2. Attempt any **three** of the following: **3x5=15**

(a) What are the major causes of deforestation? Discuss its consequences.

(b) What is biogas production? Explain the anaerobic digestion process.

(c) What do you understand by conventional and non-conventional energy sources?

(d) Enumerate major categories of pollution of environment. What is their general fate?

(e) What do you understand by accelerated urbanization? Discuss its consequences.

Section C

3. Attempt any **five** questions: **5x5=25**

(a) Discuss the climate change and global warming. Also describe some control measures.

(b) What are the causes of ozone layer depletion? Explain its consequences.

(c) How and why we can say that hydrogen as an alternative future source of energy?

(d) Use of toxic chemical and radioactive material has added a new public health challenge? Discuss.

(e) What are acid rains? Discuss their causes, consequences and control measures.

(f) What do you understand by sustainable development? Trace the emergence of global concern for sustsinable development.

(g) Write short notes on any **two** of the following:

(i) Animal Husbandry

(ii) Water borne diseases

(iii) Economic and Social security

B. Tech (Sem I) Theory Examination, 2010-11, Environment and Ecology Paper ID 3103

Time: 2 Hours] *[Total Marks: 50*

Note: Attempt **all** questions from all sections as per instructions.

Section A

1. Attempt all questions which are compulsory. **10x1=10**

(a) A World Environment Day is observed on______.

(b) Chipko movement was started to conserve ______.

(c) Acid rain is caused due to emission of ______ to the atmosphere.

(d) Balika Samridhi Yojna is a measure of ______ welfare.

(e) The ecological pyramid which is always straight is ______.

(f) The maximum value of requirement (desirable limit) of fluoride (as F) in drinking water as per IS 10500 is ______.

(g) Gas leaked in Bhopal Tragedy was ______.

(h) The study of interaction between living and non-living organism and environment is called as ______.

Indicate True or False for the following statements:

(i) Algae is an example of Protozoa (True/False).

(j) Green marketing and eco-labelling are good strategies to protect environment from waste products. (True/False).

Section B

2. Attempt any **three** of the following: **5x3=15**

(a) Write an explanatory note on the multidisciplinary nature of environmental science.

(b) Discuss the environmental effects of extracting and using mineral resources.

(c) Write down National Ambient Air Quality Standards for industrial, residential, rural and other area as prescribed by Central Pollution Control Board.

(d) Discuss the Indian scenario in solar energy utilization and development.

(e) Explain intensity, power and pressure levels of noise.

Section C

Note: Attempt any **five** questions: **5x5=25**

(a) Define ecology and ecosystem. Explain the role of producers, consumers and decomposers in an ecosystem.

(b) With the help of flow chart, describe the Environment Impact Assessment process.

(c) What is deforestation? Enumerate and discuss the various causes of deforestation.

(d) Explain the sulphur cycle with neat sketch.

(e) (a) Define air pollution. What are the sources of air pollutants? How will you classify air pollutants.

(b) What are the sources and effects of solid waste? Explain waste minimization techniques.

(f) (a) What are greenhouse gases? Name and discuss their contribution to global warming. What can be the effects of global warming?

(b) The Earth system is not and never has been free from climate change' Comment.

(g) (a) What is ozone hole? What are the causes of ozone hole formation? What are the effects of depletion of ozone layer?

(b) What should be the 'guidelines of excellence' for environmental education?

(h) (a) Discuss the strategy and policy adopted by Government of India for the development of women education?

(b) Discuss the role of Government and legal aspects in environmental protection.

(i) Write short notes on following:

(a) Effect of transportation activities on environment.

(b) Hydrogen as an alternate future source of energy.

B. Tech (Sem II) Theory Examination, 2007-08, Environmental Studies Code: TES 201

Time: 3 Hours] *[Total Marks: 100*

Note: Attempt **all** questions from all sections as per instructions.

Section A

1. Attempt any **four** parts of the following. **5x4=20**
   (a) How would environmental awareness help to protect our environment?
   (b) How would you broadly divide the major layer or regions of atmosphere? State their respective altitudes and temperature ranges.
   (c) Discuss with the help of a case study, how big dam is affected forest and the tribal?.
   (d) Discuss the environmental effects of extracting and using mineral resources.
   (e) What is nuclear energy? Discuss its two types.

2. Attempt any four parts of this question. **5x4=20**
   (a) Define ecosystem. Give an account of the structure and function of an ecosystem.
   (b) Fuel powered ecosystem are man's wealth generating and also pollution generating system Justify.
   (c) Define ecological succession. Give an account of general process of ecological succession.
   (d) What are the essentials of the desert ecosystem?
   (e) What do you mean by Thermocline and Oligotrophic lakes?
   (f) Write a note on tropical rain forest and savannas.

3. Attempt any two parts of the following: **10x2=20**
   (a) What do you understand by the term "Biodiversity"? Name and discuss the three hierarchical levels to study the concept of biodiversity.
   (b) What do you understand by 'conservation of biodiversity'. State and explain the two basic approaches of wild-life conservation.
   (c) Define the threats to Biodiversity and also explain endangered, threatened and endemic species of India.

4. Attempt any two parts of the following: **10x2=20**

(a) Enumerate the various natural mechanisms that work in atmosphere. Which is the most important particulate removal mechanism?.

(b) Breifly describe the composition of domestic and industrial waste water. What are the basic differences between them?

(c) What is recycling? Discuss its importance, by giving suitable examples, in solid waste management. Also give advantages of recycling and waste utilization.

5. Attempt any four parts of the following: **5x4=20**

(a) What is rainwater harvesting? What are the purposes served by it?

(b) Discuss the natural formation and occurrence of ozone on the stratosphere.

(c) What is meant by population explosion? Discuss the Indian scenario.

(d) What are the objectives and elements of value education? How can the same be achieved?

(e) Discuss El-Nino phenomenon and its effects at global scale.

(f) How can we use IT in rural areas? Explain.

B. Tech (Sem II) Theory Examination, 2005-06, Environmental Studies Code: TES 201

Time: 3 Hours] *[Total Marks: 100*

Note: Attempt **all** questions from all sections as per instructions.

Section A

1. Attempt any **four** parts of the following. **5x4=20**

(a) With the help of neat sketch discuss the structure of atmosphere along with temperature profile of atmosphere and related phenomenon.

(b) Unawareness or ignorance with protection of environment will lead to detrimental consequences-comment with illustrations.

(c) What are the different types of minerals? Describe the effects of mineral extraction on environment.

(d) Name and explain in brief, the various ways in which, energy from oceans can be obtained.

(e) Describe geothermal energy. Explain its merits and limitations.

(f) Write short notes on the following:

(i) Environmental movement in India

(ii) Environment related acts.

2. Attempt any **two** parts of this question. **10x2=20**

(a) Depict diagrammatically a terrestrial food web and oceanic food web.

(b) Describe the structure and function of the forest ecosystem and lake ecosystem.

(c) "The flow of energy is one-way and continuous in an ecosystem" Justify.

3. Attempt any **two** parts of the following: **10x2=20**

(a) Enumerate different biogeographic zones of India and explain the characters of each.

(b) What do you understand by ' Hotspots of Biodiversity'? Name and briefly describe along with vital signs, the two hotspots of biodiversity that extend into India.

(c) Write short notes on any **two** of the following:

(i) Rare species and threatened species

(ii) Poaching of wild-life

(iii) *In situ* conservation of biodiversity.

4. Attempt any two parts of the following: **10x2=20**

(a) What do you mean by water pollution? Discuss its major sources and effects. Discuss in brief various processes of water treatment with layout of water treatment plant.

(b) What are the causes of landslides? Discuss the consequences and suggest the methodology to fight bach such hazards.

(c) What are solid wastes? Discuss their types, effects and name various methods used to dispose solid wastes. Explain any one of them with their merits and demerits.

5. Attempt any four parts of the following: **5x4=20**

(a) What is rain water harvesting? Name and discuss in brief the types of rain water harvesting.

(b) Briefly discuss the objectives of resettlement and rehabilitation policy.

(c) What is ozone-hole? What are the cause of ozone hole formation. Discuss the effects of ozone layer depletion and its remedial measures.

(d) Briefly discuss the salient features of the Environment (Protection) Act, 1986.

(e) What is ' value education'? Discuss the concept of value education with suitable examples.

(f) What do you understand by the term ' Human Rights'. List the important articles of the 'Declaration of Human rights, 1948' adopted and proclaimed by United Nations.

B. Tech (Sem II) Theory Examination, 2004-05, Environmental Studies Code: TES 201

Time: 3 Hours] *[Total Marks: 100*

Note: Attempt **all** questions from all sections as per instructions.

Section A

1. Attempt any four parts of the following. **5x4=20**

   (a) Discuss briefly the impact of over utilization of water on ground and surface water.

   (b) What do you meant by deforestation? Explain its causes and ill-effects.

   (c) What are the alternative energy resources? Differentiate between renewable and non-renewable natural resource.

   (d) Name and discuss the tqo major world food problems.

   (e) Discuss the energy scenario in Indian context.

   (f) Discuss the environmental effects of extracting and using mineral resources.

Section B

2. Attempt any **two** parts of this question. **10x2=20**

   (a) What is an ecosystem? Describe briefly the forest ecosystem and aquatic ecosystem.

   (b) Define food chain. Name and explain various types of food chains with suitable examples.

   (c)

   What is ecological succession? Describe the causes and basic types of ecological succession.

Section C

3. Attempt any two parts of the following: **10x2=20**

   (a) What is meant by Biodiversity? What is its concept; describe the uses and importance of biodiversity.

   (b) Write an explanatory note on the threats to biodiversity.

   (c) Explain *in-situ* and *ex-situ* approaches of conservation of biodiversity. Compare their advantages and limitations.

4. Attempt any two parts of the following: **10x2=20**

   (a) What is air pollution? What are its effects on human health? Discuss the

(b) What do you understand by natural disasters? What measures you would take to prevent loss of lives from floods and earthquakes? Explain.

(c) Explain the various methods commonly employed for the disposal of solid wastes, with their advantages and disadvantages.

5. Attempt any four parts of the following: **5x4=20**

(a) What are wastelands? Name and discuss the various methods of wasteland reclamation.

(b) Discuss population explosion in Indian context. What should be the objectives of a sound population policy?

(c) What is AIDS? What are the sources and mode of transmission of HIV infection?

(d) Briefly discuss the salient features of the Forest (Conservation) Act, 1980.

(e) Discuss the significance of rain water harvesting and watershed management.

(f) Discuss briefly the environmental ethics and related issues with their solutions.

Modal Paper as per UPTU Pattern

ENVIRONMENT and ECOLOGY EAS-105, 205 PAPER ID 3103

[Total Marks: 50

Section A

1. Fill in the following blanks with suitable words: **(10x1=10)**

   (a) International conference for reducing CFCs is ______, for reducing GHG is ______ and ______.

   (b) EPA, 1986 is the most ______ environmental law. Our environment minister is ______.

   (c) NIDM stands for______. CSE stands for ______. NEERI stands for ______. CEE stands for ______.

   (d) Pyramid of number of parasitic food-chain is ______.

   (e) Unit of measurement of earthquake is ______. Unit of ozone is ______. Unit of noise is ______.

   (f) Acid Rain is a ______ Pollutant.

   (g) The term ecosystem is given by ______. The study of ecosystem is known as ______.

   (h) The study of electromagnetic radiation is called as ______.

   (i) No-go zones are the dense forest cover where mining is permitted by Jai Ram Ramesh (True/False).

   (j) Ozone hole is found above the Antartica on the south-pole. Ozone hole is a single reaction of CFCs with Ozone (True/False).

Section B

2. Attempt any three parts. All parts carry equal marks: **(5x3=15)**

   (a) Discuss the role of Government and legal aspects in environmental protection.

   (b) Discuss the environmental effects of extracting and using mineral resources. Also explain the detrimental effects on forests.

   (c) What are the causes, effects and control measures of Water Pollution? Describe the water quality standards given by CPCB.

(d) Describe the significance of Non-conventional energy resources. Discuss the Indian scenario in solar energy utilization and development.

(e) What is 'Sustainability'? Explain the principals and strategies of Sustainable Development. Or Describe the concept of Journey from Urbanization to Sustainable Development.

Section B

3. Attempt any Five questions **(5x5=25)**

3. (a) Discuss the concept of balanced ecosystem. Comment on Energy Flow. Draw the energy flow model for lake ecosystems (Write all units)

(b) With the help of flow chart, Describe role of Environment Impact Assessment in engineering projects.

(c) Describe the ill-effects of modern agriculture or explain fertilizer pesticide problem.

4. (a) Write an explanatory note on Fluorosis. Or Comment "Fluorine is both good and bad for health".

(b) Explain Nitrogen cycle with neat sketch. List the processes and microorganisms involved.

5. (a) Define noise pollution. What are the sources and effects of noise pollution?

(b) What are the sources and effects of solid waste? Explain waste minimization or management techniques.

6. (a) What are greenhouse gases? What can be the effects and control measures of global warming?

(b) Discuss the adverse effects of nuclear explosions occurred recently in Japan. Is the disaster can be co-related with Hiroshima Nagasaki case-study. Suggest the solutions.

(c) Write about the science of animal-keeping. Its benefits and environmental concerns.

7. (a) What is Acid Rain? Explain the causes, effects and control measures.

(b) Comment upon the occurrence of earthquake in Japan. How would you propose safety-measures.

(c) Describe urbanisation and its advantages and disadvantages.

8. (a) Discuss the role of women education in sustainable development. What are the problems of women education and policies adopted by government.

(b) Explain Environmental Education. Discuss the role of NGOs in environmental protection.

9. (a) Define Natural Resources. Explain the types and their importance.

   (b) Describe energy scenario with respect to Indian perspective.

Note: ALL QUESTIONS ARE COMPULSORY. Time : 2Hrs

Q1: Fill in the blanks. **1X10=10**

(a) Biogas predominantly contains ______.

(b) The pyramid of ______ is always upright or straight.

(c) The major responsible greenhouse gas is ______. It contributes______(per cent) to global warming.

(d) Balika samridhi yojana is related to ______.

(e) PAN is formed due to reaction with ______ and ______.

(f) Due to high concentration of mercury in the water body, the disease caused is ______and the process is known as ______.

(g)

Narmada Bachao Andolan was initiated by ______to rehabilitate people, displaced due to ______ project.

(h) Rainwater harvesting is a tool to Water Shed Management.

Q2: Tick, True or False

(i) BOD is amount of dissolved oxygen required by microorganisms to degrade organic matter.

(j) Tropospheric ozone is a pollutant.

Q3: Answer any five of the following. **2X5=10**

(a) Describe food-chain and explain its type.

(b) Describe food related problems in the biosphere.

(c) Write Short note on Fluoride problem in drinking water and its control measures.

(d) Describe radioactive or nuclear pollution with the help of recent happening's example.

(e) Write short-notes on eutrophication and blue-baby syndrome (methaemoglobinemia).

(f) Define lapse rate. Describe how it varies in atmosphere.

Q4: Answer any five parts of the following: **5X6=30**

(a) Explain the salient features of Environmental Protection Act, 1986 and

(b) Comment "Energy flow is unidirectional and continuous". Describe the structure and function of grassland and pond ecosystem.

(c) Describe the concepts of Sustainable Development and Environmental Impact Assessment.

(d) Explain the type of Energy Resources. Describe two energy resources from each type.

(e) Comment "Open dumps are causing environmental hazards" Describe how to resolve.

(f) Define the following:

- ☆ Green House Effect
- ☆ Global Climate Change
- ☆ Ozone layer depletion
- ☆ Acid Rain
- ☆ Animal Husbandry

(g) Differentiate Conventional, Non-Conventional and Perpetual energy resources. Describe advantages and disadvantage of following:

- ☆ Geothermal Energy
- ☆ Nuclear Energy and Windmills

Glossary

Abiotic Non-living or physical components of environment.

Acid Rain or Acid Precipitation or Wet Precipitation (Stone Leprosy) Oxides of sulphur, nitrogen and oxygen are transformed to acid and deposited on the earth surface in wet or dry form. In wet form, it is known as Acid Rain or Acid Precipitation. The dry forms are acidic gases or particulates known as Dry Precipitation.

Activated Sludge Semisolid waste of treatment plants resulting when the primary effluent is mixed with bacteria-laden sludge and then agitated and aerated to promote biological treatment to enhance the rate of degradation of organic matter.

Adaptation The small and continuous evolutionary changes in the physiology of the living organism in order to suit the abiotic conditions.

Aerosol Small droplets or particles (preferably constituting sulphur) suspended in the atmosphere, emitted from volcanic eruptions and burning of fossil fuels, are known as Aerosols.

Afforestation Planting trees on bare land *i.e.* Converting bare land to forest or Planting new trees.

Agroforestry Sustainable agriculture to plant trees in the farm to enhance socio-economic and environmental benefits. Fruit-giving trees are planted at the corners and centre of agricultural land.

Air Contaminant Any foreign body (liquid or gaseous, biotic) present in the atmosphere which degrades the quality of air.

Air or Atmosphere Mixture of gases and water-vapour comprises air.

Air Pollutant Any substance present in air in concentrations higher than their carrying capacity in the air, causes adverse effects in the physical, chemical and biological properties of air and is harmful to life.

Air Pollution The phenomenon due to which the presence of pollutants in the air, causes adverse effects in the physical, chemical and biological properties of air and is harmful to life.

Algal Bloom Excessive growth of algae due to nutrient enrichment (fertilizer intake) of water-body, stops inter-mixing of air with water.

Alternative Fuels To reduce the pressure on Non-renewable energy resources and to reduce air pollution, ethanols and Compressed Natural Gas have been used as potential fuels.

Ambient Air The air which immediately surrounds us.

Anaerobic Decomposition Breakdown of complex substance to simpler form, in the absence of oxygen.

Animal Husbandry The branch of science, which deals with rearing of domestic animals, under human care.

Aquifer A bed of rocks, sand, gravel or other porous materials yielding water.

Atmospheric Nitrogen Fixation When lightening takes place in the atmosphere, Nitrogen in the atmosphere reacts with the Oxygen molecules and is converted to usable form by plants (Nitrate).

Automobile Pollution The gases emitted from the exhaust of the engine of automobiles is known as automobile pollution.

Autotrophs Green plants make their own food in the presence of CO_2, water and Sunlight by the process of photosynthesis.

Balanced Ecosystem Energy and nutrient flow is maintained in an ecosystem to form a steady state mechanism in an ecosystem known as Homeostasis.

Bioaccumulation The process of selective absorption of certain molecules in the cell.

Biochemical Oxygen Demand The amount of dissolved oxygen consumed by the microorganisms in the process of degradation of organic matter present in the water.

Biofertilizers The nutrients obtained from living organisms *e.g.* Alfaalfa is nitrogen rich plant used as biofertilizer.

Bio-fuels The fuels produced by microbial activity with organic matter.

Bio-gas Mainly methane, obtained by anaerobic decomposition of organic matter.

Biogeochemical Cycle (Biogentic Nutrient Cycle or Nutrient Flow or **Nutrient Cycling**) Nutrients keep flowing in the various environmental segments (Air, water, soil, biotic components) via food-chains and food-webs. It is bidirectional or cyclic in nature.

Biomagnification Increase in concentration of stable toxic chemicals, with increasing trophic level in a food-chain.

Biomass Energy Energy obtained from biomass.

Biomass The total weight of living organisms in a given population of an area.

Biome A broad terrestrial region, marked by distinctive vegetation and climatic conditions

Biopesticides Certain micro-organisms are used to kill the pests which adversely affect the growth of pests.

Bioremediation To remove pollutants from air, water and soil, living organisms are used. *e.g.* Air pollution tolerant trees absorb polluted air *i.e.* SO_X, NO_X, CO_X.

Biosphere The integrated part of air, water and soil where life exists.

Biotic Community A group of populations of different species interacting with each other living in an ecosystem.

Biotic Component The living component of an ecosystem.

Carnivores (Secondary consumers) Organisms that eat other consumers. *e.g.* fox.

Carrying capacity Maximum number of individuals, that can be supported in an ecosystem.

Chemical Oxygen Demand The amount of dissolved oxygen required by chemicals to breakdown organic and inorganic matter, in water.

Chlorofluorocarbons (CFCs) A group of chemical compounds with carbon skeleton and attached Chlorine and Fluorine atoms, widely used in refrigerators, air conditioners, packaging, insulation, as solvents and aerosol propellants.

Chlorosis Degradation of green chlorophyll due to disease, lack of nutrients and air pollution.

Clean fuels Fuels which emit less air pollution as compared to traditional non-renewable fuels (coal). *e.g.* biofuels, bioethanol, biogasohol, syngas *etc.*

Climate A single term to explain all the abiotic conditions, in an ecosystem, representing average weather conditions (general pattern of atmospheric seasonal variations and weather extremes.

Climate Change Change in weather conditions (temperature, rainfall, humidity *etc.*) over a long period of time, due to global warming.

Climax The stable stage of ecosystem achieved in the process of ecosystem development or ecological succession.

Community The population living in an area.

Competition When two species interact with each other, they compete for sunlight, air, water and nutrients.

Composting Nutrient rich soil conditioner formed by degradation of organic matter to simple form.

Compressed Natural Gas (CNG) CNG is made by compressing natural gas which is mainly composed by methane (CH_4).

Confined Aquifer Aquifer between two impermeable layers of earth.

Conservation Tillage Crop farming where soil is disturbed to least.

Consumerism An attitude attracted towards consumption of natural resources.

Contaminant Any foreign body (liquid, air or gaseous, biotic) present in the environment degrades the quality of air, water or soil, causing harmful effect to life.

Contamination The phenomenon due to which any foreign body (liquid, air or gaseous, biotic) present in the environment degrades the quality of air, water or soil, causing harmful effect to life.

Contour Farming Farming perpendicular to the slope of an area to reduce soil erosion and water loss.

Decibel Physical unit for measurement of sound.

Decomposers A group of microorganisms which breakdown (eat) complex dead organic matter to simpler form (also known as Saprotrophs, Detrivores, Reducers).

Deforestation Cutting of trees, enhances soil erosion, disturbs water-cycle, perturbs habitat of many animals, reduction in availability of medicines, fruits, gums, resins and other Non-Timber Products of forest *etc.*

Denitrification The process of conversion of nitrates and nitrites to atmospheric nitrogen *e.g. Pseudomonas* is a denitrifying bacteria.

Desert A biome where high temperature and very less rainfall, causes scarcity of water.

Desertification Conversion of forest or grassland into desert.

Detritus Dead organic matter.

Detrivores Microorganisms that feed upon dead organic matter *e.g.* insect larvae.

Dioxins A group of chemical compounds, highly toxic to living organisms, formed in industrial operations, when waste plastic, coal and cigarettes are burnt.

Dissolved Oxygen The amount of oxygen dissolved in water, available to plants and animals for respiration.

Dobson Unit (DU) 1DU=1ppb, Thickness of ozone is measured in Dobson units.

Doubling Time Time required for a population to double in number.

Drip Irrigation For watering crops, perforate tubes are used, to supply water drop-wise, to avoid wastage of water.

Dry Deposition Gases and particles settle down on the earth surface and cause adverse effects.

Earthquake Shaking of ground due to fracture and displacement of plates of earth crust.

Ecological Footprint The concept given by scientists of Netherlands which refers to the area of earth needed by an entity to sustain its life.

Ecological Niche The physical, chemical and biological factors of the natural habitat of a species, where it naturally lives and reproduce.

Ecological Succession The very slow process of ecosystem development, in which, living organisms undergo adaptations and keep evolving.

Ecology The interrelationship between living and non-living factors is known as ecology.

Ecosphere The area where life exists *i.e.* integration of atmosphere, lithosphere and biosphere.

Ecosystem goods Goods provided by ecosystem as a result of mutual interactions between living and non-living factors.

Ecosystem Living and non-living factors interact with each other to form a self-sufficient unit.

Ecosystem Services Services provided by ecosystem as a result of mutual interactions between living and non-living factors. *e.g.* pollination, water-cycle, soil development, air purification *etc.*

Edaphic Factors related to soil.

Effluent The waste emitted from industries to air or water bodies *e.g.* smoke and waste water.

El-nino A phenomenon due to climatic change wherein shifting of a large warm water pool from the western Pacific Ocean towards the east. Wind direction and precipitation patterns get changed over the Pacific and around the world.

Energy Plantation Plantation of crops on large scale, which provide energy.

Energy Recovery Extraction of energy from waste material.

Environment The sum total of physical, chemical, biological, economical and social factors that contribute to the environment.

Environmental Audit It is an inspection exercise in any industry to ensure its usage, processes followed in compatible with environmental conservation.

Environmental Impact Assessment (EIA) The continuous process, in which proponent of mega-developmental project analyses the pros and cons of a developmental project, minimizes the negative environmental impacts, thus helps in sustainable development.

Environmental Impact Statement (EIS) Report summary of EIA.

Environmental Pollutant or Pollutant The agent present in the air, water and soil causes adverse effects in the physical, chemical and biological properties of air, water and soil and is harmful to life.

Environmental Pollution or **Pollution** The phenomenon due to which the presence of pollutants in the air, water and soil causes adverse effects in the physical, chemical and biological properties of air, water and soil and is harmful to life.

Environmental Sciences The branch of science, which analyses interactions between living and non-living factors and role of man in its surroundings. It is an integration of life sciences, social sciences, physical and chemical sciences, ecology, engineering and management.

Environmental Studies The branch of study which deals with environmental issues. It covers broader aspects of environmental sciences.

Estuary The water-body wherein rivers meet at a point in oceans *i.e.* it is a combination of freshwater and saline water *i.e.* it is rich in biodiversity of both the regions.

Eutrophication Excessive growth of algae due to nutrient enrichment (fertilizer intake) of water-body, stops inter-mixing of air with water, causing death of water plants and animal, putting water-body under threat. Such a water-body is known as eutrophic water body and the process is known as Eutrophication.

Evapotranspiration Evaporation of water from the leaf surface.

E-waste Electronic waste generated from the waste of discarded electronic devices *e.g.* computers, televisions, telephones, mobile phones, music systems *etc.*

Exosphere The uppermost layer of atmosphere in which, continuous increase of temperature occurs with height.

Fluorosis Disease caused due to excessive intake of fluoride in the drinking water, interferes with the calcium metabolism. It is of two types: skeletal fluorosis (weakening of bones) and dental fluorosis (weakening of teeth).

Flyash Solid waste generated due to incomplete combustion of coal.

Food-chain Transfer of food energy from one organism to another organism, in an ecosystem.

Food-web An internetworking of foochains, to maintain balance in an ecosystem.

Fossil Fuels Non-renewable energy resources, which take a very long passage of time, in their formation inside the earth, when dead plants and animals are buried at very high temperature and pressure.

Freons Trade name of CFCs which cause ozone layer depletion.

Fuel Cell An electrochemical cell that burns hydrogen to generate electricity.

Fumes The smoke or plume emitted from the industry contains gases and particulate matter.

Fungicide The chemical, which kills the fungus, infecting the crop plants.

Geographic Information System (GIS) Sensor based mapping of an area to collect wider information on sensitive issues *e.g.* weather forecasting,

Global Warming Drastic rise in the temperature of earth, due to emission of large amount of Green House Gases *e.g.* CO_2, CH_4, NO_2, N_2O, SO_2

Greenbelt Designing Planting air pollution tolerant trees in an area around industries, where air pollutants are predicted to precipitate (fall) on earth.

Greenhouse Effect The process of absorption of large amount of sun-rays (Infra-red radiations), by the Green House Gases.

Greenhouse Gas The gas that captures heat rays in the atmosphere *e.g.* CO_2, CH_4, NO_2, N_2O, SO_2.

Groundwater Water held below the earth surface in the pores of rocks.

Gully Erosion Removal of large layers of soil, creating channels difficult to be removed by normal tillage operations.

Habitat The place of conditions where an organism naturally lives.

Halons Group of chemicals commonly used in fire extinguishers. They cause ozone layer depletion because they release bromine atoms when released in the atmosphere.

Hazard The phenomenon which causes versatile adverse impacts to life *e.g.* natural hazards

Hazardous Waste The waste which is lethal to life *e.g.* cyanide.

Heap-leaching or Heap-leach Mining Process of gold mining, in which cyanide is poured on huge pile of gold rock from which metal is extracted and contaminates ground water with cyanide.

Herbivores or Primary Consumers Organisms that eat directly on green plants.

Hydrological Cycle (water-cycle) The natural process of evaporation of water, its condensation in the air and precipitation of water to the earth surface.

Hydrology The study of water component of the earth.

Integrated Pest Management (IPM) The sustainable agriculture, which integrates traditional and modern agricultural techniques of pest control *i.e.* optimum use of biopesticides and chemical pesticides.

International treaties and convention Agreement signed by different nations.

Jhoom cultivation (Splash and burn agriculture) Ancient farmers follow the practice of felling and burning of crops, once the cropping is one on a farmland and then practicing agriculture on another farmland.

Joint Forest Management Sustainable management of forest's resources by the local people in the forest's conservation programmes.

Kyoto Protocol An international agreement to reduce greenhouse gases emission.

Lanslides Massive rock movements on the earth surface.

Leachate The liquid which moves downward along with water.

Leaching of toxicants Downward movement of toxicants dissolved in water.

Lethal The things which are so harmful to life that causes immediate death.

Lithosphere The solid dry earth crust that governs continents and other land masses.

Macrophytes Rooted plants which grow in shallow water.

Methaemoglobinemia It is caused by decreased ability of blood to carry oxygen, resulting in oxygen deficiency in different body parts.

Microclimate Climatic conitions in the small area.

Mine spoil (Delerict Land) Wasteland caused due to mineral extraction.

Monoculture Rearing of a single type (species) of crop or tree.

Montreal Protocol An international agreement to reduce the emission of ozone depleting substances (by finding substitute of CFCs).

Mortality (Crude Death rate or Death rate) The number of deaths occurred per thousand individuals, in a population in a given year.

Municipal Solid Waste The waste which is generated from households in known as MSW.

Natality (Crude Birth rate or Birth rate) The number of new borns per thousand individuals, in a population in a given year.

Necrosis Killing of plant tissues due to air pollution, disease and deficiency of nutrients.

Nitrification The process of conversion of Ammonia to Nitrite and to Nitrate by nitrifying bacteria.

Nitrogen Cycle Movement of nitrogen (in gaseous or compound form) among the various environmental segments.

Nitrogen Fixation Conversion of atmospheric nitrogen to a form, which plants can absorb *i.e.* nitrates, ammonia and protein.

Nitrogen Fixing Bacteria Certain micro-organisms present in the root nodules of leguminous plants absorbs atmospheric nitrogen and fixes

Noise The unwanted sound which cause adverse environmental impacts.

Non-point Source Pollution The sources of pollution which keeps moving *e.g.* automobiles.

Nuclear Fission Heavy nuclei (isotopes with large mass number) on bombardment with neutrons split into lighter nuclei, releasing large amount of energy.

Nuclear Fusion Two lighter nuclei (isotopes with small mass number) fuse together to form a heavier nucleus, releasing very large amount of energy.

Nutrient Cycle The movement of nutrients in the environmental segments *i.e.* air, water, soil.

Oil Spills The accidental leakage of oil in the sea during its transport.

Open-Dump The waste is thrown in low-lying areas in the outskirts of the city, in most of the cities in India, causes environmental pollution.

Organic Farming Sustainable agricultural practices based on use of biofertilizers and biopesticides, to obtain safe good quality yield, while maintaining the soil fertility.

Overgrazing The removal of fertile grassland due to excessive grazing by animals.

Ozone Depletion The thinning of ozone layer in the stratosphere, due to excessive use of ChloroFluoroCarbons (CFC), in refrigerators, air-conditioners, aerosols, and rocket propellants.

Ozone Hole The thinning of ozone layer in the stratosphere, due to excessive use of ChloroFluoroCarbons (CFC), in refrigerators, air-conditioners, aerosols, rocket and propellants.

Ozone Layer A 8 ppm thick layer of ozone molecules is present in the stratosphere, absorbs harmful Ultra-violet radiations.

Ozone The gaseous molecule, molecular formula O_3.

Pathogens The disease causing micro-organisms.

Pest The insect which harms the crop.

Pesticide The organic or inorganic compound which kills the insects which harm the crop.

pH Negative logarithm of Hydrogen ions concentration.

Phosphorus It is a macro-nutrient required by plants for their growth.

Photochemical Smog A smoke formed as a result of photochemical reaction between the products of incomplete fuel combustion and oxides of nitrogen. It is composed of O_3 (Ozone), PAN (Peroxy Acetyl Nitrate) and N_2 (Nitrogen).

Photosynthesis The natural process, by which green plants make their own food in the presence of CO_2, water and sunlight.

Photovoltaic Cell (PV cell or solar cell) Semiconductors convert solar energy into electricity.

Phytoplankton The very small water plants *e.g.* algae.

Plankton Very small microscopic plants and animals suspended in water and being eaten up by animals.

Population A group of individuals of same species living in the same area.

Population Explosion The exponential growth in population puts pressure on natural resources, causing scarcity of natural resources.

Predation The process in which an organism (the prey) is killed and eaten by the other organism (predator).

ppm The physical unit to measure the concentration of a substance where x parts of a substance present in 999,999 parts of other material is x ppm.

Picocurie (pCi) or microcurie The physical unit to measure radioactive disintegration, is equivalent to $3.7 x 10^{-2}$ disintegrations per second.

Suspended Particles Substances that do not settle in the atmosphere due to their small particle size.

Thermal Pollution The change in water temperature of a water-body, where industrial hot wastewater is discharged.

Thermosphere The uppermost layer of atmosphere in which, continuous increase of temperature occurs with height.

Temperature Inversion When temperature decreases with height, due to atmospheric phenomena or air pollution.

Threatened Species Species whose individuals are still abundant in number, but need conservation efforts and if not saved may become extinct.

Tolerance Limits The range of environmental factors, within which an organism can survive and reproduce.

Toxicity Ability of a substance to cause death or illness.

Troposphere The lowest layer of atmosphere, just above the earth surface. Environmental pollution and all weather-events are caused in the troposphere.

Wetland A land area submerged with groundwater and surface-water. Vegetation that grows in wetlands is adapted to saturated water conditions.

Rem The physical unit to measure radiation energy is roentgen equivalentman (rem).

Precipitation Rainfall.

Primary Pollutants Pollutants that enter the environment, directly from the source of pollution.

Primary Treatment of Waste Water The first treatment methods to purify waste water, which are easy in operation.

Producers The green plants make their own food by the process of photosynthesis, in the presence of CO_2, water and sunlight.

Productivity The rate of production of biomass by the organisms.

Pyrolysis The burning of non-biodegradable toxic waste under high temperature and pressure.

Radiatively Active Gases Green House Gases.

Rainwater Harvesting Collection and storage of rainwater to maximize its use.

Reclaimation or Restoration Regaining the quality of land before developmental activity or wasteland reclaimation.

Reforestation Planting trees in the area where they have been cut.

Remote Sensing Satellite based sensing technique to analyse the large scale information.

Renewable Energy Source Source of energy which are renewed or formed at a faster rate by natural processes and hence can be used for longer period, to ensure sustainable future.

Sacred Grove A forest patch which is worshiped and protected by local community.

Scare-Crow Dummy hanged in the farms by ancient farmers to protect the crop from pet attack.

Sludge The semi-solid toxic material left after the treatment of sewage.

Smog A secondary air pollutant formed by combination of smoke and fog.

Social Forestry Planting trees by local people on the wasteland.

Soil Erosion Removal of top fetile layer of soil by man, air and water.

Solid Waste Rejected matter in solid form, which cause environmental pollution when disposed off.

Species Extinction The loss of certain species of organisms due to man-made or natural causes. Man-made causes of species extinction are hunting, habitat loss *etc*. Natural causes are disease, infection *etc*.

Species The similar group of individuals which can interbreed freely among themselves. *e.g.* species of rabbit, deer, rose, mango, human beings.

Threatened Species Species whose individuals are still abundant in number, but need conservation efforts and if not saved may become extinct.

Total Suspended Solids Substances that do not settle in the atmosphere due to their small particle size.

Sustainable Development The developmental activities carried out in such a way to integrate social, economic and environmental benefits.

Tertiary Consumers Organisms that feed upon other carnivores.

Threshold The upper limit or carrying capacity.

Topograpgy The nature of land area *i.e* valley, mountains, plains.

Total Fertility Rate The average number of infants born to a woman in her lifetime.

Toxicant The substance which is harmful to the living component of the environment.

Toxicity The property due to which the substance causes harmful effects to the living beings.

Transpiration Evaporation of water from the leaf surface.

Troposphere The closest atmospheric layer to the earth surface extending upto 14 kms.

UNCED (United Nations Conference on Environment and Development or Earth Summit) International conference held in Rio de Janerio in 1992.

UNCSD (United Nations Commission on Sustainable Development) An United Nations organization which continuously reviews progress in the implementation of outcomes of the earth summit held in Rio de Janerio in 1992.

Undernourishment Lack of availability of food mainly due to poverty.

Urban Forestry Technique of growing ornamental and fruit trees in urban land areas.

Urbanization Development of new town, providing social, economic and environmental benefits.

Water Pollutant The presence of pollutants in the water causes adverse effects in the physical, chemical and biological properties of water and is harmful to life.

Water Pollution The phenomenon due to which the presence of pollutants in the water causes adverse effects in the physical, chemical and biological properties of water and is harmful to life.

Waterlogging The condition of rise in water-table, due to which the soil air is replaced by soil water, causing less availability of oxygen (air) to the roots of the plants to respire.

Watershed Management Management of air, water and soil resources in such a way that efficient utilization of the rain water and its runoff ensured.

Watershed The area which captures rain.

Water-table Ground-water bodies.

Weather Short term properties of atmosphere at a given time and place.

Wetland Terrestrial ecosystem with large amount of water accumulated in the area and having rooted vegetation.

Wetlands Biomes with standing water in the soil and rooted vegetation.

Temperature Inversion When temperature increases with height, due to atmospheric phenomena or air pollution.

WSSD (World Summit on Sustainable Development) held in Johannesberg, South Africa wherein the international agenda for sustainability is think globally and act locally. Poverty alleviation, environmentalconservation are the major issues.

Zero Population Growth When birth rate equals the death rate and immigration rate of population equals the emigration rate.

Zoo The place where animals are kept under human care *i.e.* rearing of animals and providing opportunities to interbreed, to increase their number. Thus it is a *ex-situ* conservation of biodiversity.

Zooplankton Very small animal like structures *e.g.* insect larvae.

Index

A

Abiotic Components 12
Acid Rain 182
Activated Sludge Process 145
Aerobic Treatment 144
Agroforestry 67
Air (Prevention and Control of Pollution) Act, 198 168
Air Pollution 131
Air pollution 41
Air-borne Diseases 155
Alpha-diversity 80
Animal Husbandry 202
Arsenecosis 65
Arsenic 42
Automobile Pollution 174

B

Balanced Ecosystem 9
Bauhinia variegate 41
Beta-diversity 80
Bhopal Gas Tragedy 136
Biodiversity 74
Biodiversity Management 90
Biodiversity Treaty 94
Biofertilizers 67
Biogeochemical Cycling 31
Biological Oxygen Demand (BOD) 140
Biomagnification 54, 140
Biomethanation 149
Biopesticides 67
Bioremediation 41, 66, 140
Biosphere 4, 9
Biosphere Reserve 92
Biotic Factors 13
Black foot disease 65
Blood-brain barrier 158
Botanical Gardens 93

C

Cadmium 42
Carbon Cycle 33
Carcinogens 158
Carnivores 16
Catalytic Converters 134
Challenges to Food Security 50

Chemical Interactions 158
Chipko movement 96
Climax 40
CNG (Compressed Natural Gas) 117
Coal Energy 116
Composting 149
Crop-rotation 69
Cyclone 200
Cyclone Separator 134

D

Decomposers 16
Decontamination 42
Denitrification 36
Dental Fluorosis 64
Desert 83
Detritus Food-Chain 17
Disaster Management 198
Drought 201

E

Earth Summit 127
Earthquakes 200
Ecological Pyramids 36
Ecology 12
Ecorestoration 39, 42
Ecorestoration Technology 40
Ecosystem 8
Ecosystem Diversity 78
Ecosystem Restoration 41
Ecosystems 39
EIA Ruling 1984 163
El Nino 182
Endangered Species 89
Endemic Plants 85
Endemic Species 85
Environment (Protection) Act, 1986 172
Environmental Education 187
Environmental Factors 158
Environmental Impact Assessment (EIA) 162
Environmental Lapse Rate (ELR) or Lapse Rate 2
Environmental Laws 167
Environmental Pollution 131
Eutrophication 52, 140
Ex-situ Conservation 92

F

Farmland Conversion 55
Filters 134
Flood 199
Fluorosis 64
Flyash Landfill Reclaimation 198
Food-chain 17
Food-Web 17
Forest Conservation Act, 1980 171

G

Gamma-diversity 80
Gangetic Plains 83
Garbage Disposal 205
Gaseous Pollutant 133
Gene Bank 93
Genetic Biodiversity 76
Genetic Engineering 93
Geothermal Energy 110
GIS (Geographic Information System) 93
Global biodiversity 96
Global Climate Change 180
Global Warming 178
Grassland Ecosystem 36
Grazing Food-Chain 17
Green engineers 7
Green House Effect 178
Green manures 67
Green-chemistry 7

Green-economy 7
Green-professional 7
Ground Water Contamination 52
Growth Limiting Factors 13

H

Habitat Fragmentation 87
Hazardous Chemicals 201
Hazardous Waste Managment (HWM) 150
Health and Hygiene 204
Herbivores (Primary Consumers) 14
HIV/AIDS 209
Hotspot of Biodiversity 82
Human Population 207
Human Rights 209
Hunting 87
Hydro Energy 108
Hydrological Cycle 62
Hydrosphere 4

I

Illiteracy 204
In-situ Conservation 91
Incineration/pyrolysis 149
IUCN 93

J

Jhoom Cultivation 88
Joint Forest Management 193

K

Kyoto Protocol 128
Kyoto ProtocolAn international 128

L

La Nina 182
Land Degradation 69, 145
Landslides 58, 200
Lead 42
Lithosphere 4
London Smog 134
Los Angeles Smog 135
LPG (Liquefied Petroleum Gas) 117

M

Mangrove Ecosystem 17
Marine Ecosystems 44
Marine Energy 109
Mercury 42
Mesosphere 2
Metallothioneins 42
Microbial Remediation 44
Mixed-Cropping 67
Montreal Protocol 128
Mulching 69

N

National Ambient Air Quality Standards 136
National Park 92
National Population Policy, 2000 209
Natural Ecosystem Restoration 40
Natural Resources 61
Nitrification 36
Nitrogen Cycle 33
Nitrogen Fixation 33
Noise Pollution 58, 153
Non-Renewable Energy Resources 103
Non-Renewable Natural Resources 61
Nuclear Energy 117
Nuclear Fission 117
Nutrient Cycling 31
Nutritional status 158

O

Ocean Energy 109
Organic-Farming 67
Oxidation-pond 145
Ozone Layer Depletion or Ozone Hole 181

P

Park Conservation 91
Passive Solar House 107
Perpetual Energy Resources 103
Perpetual Natural Resources 61
Phytoextraction 42
Phytomining 42
Phytoremediation 41
Phytostabilization 41
Ploughing 69
Poaching 87
Population Explosion 100
Population Growth Rate 207
Primary Air Pollutants 132
Priority engineers 7
Process of EIA 163
Producers (Autotrophs) 14
Pyramid of Energy 39
Pyramid of Numbers 36

R

Rainwater Harvesting 198
Rare 94
Renewable Energy Resources 103
Renewable Natural Resources 61
Reverse Osmosis 145
Rhizofiltration 43
Rio +20 Earth Summit 129
River Restoration 205

S

Sanctuary 92
Scare-crow 67
Scrubber 134
Secondary Air Pollutants 132
Secondary Succession 40
Seed Suicide 54
Setting Chambers 134
Sex 156
Sexually Transmitted Diseases (STD) 209
Shifting 88
Skeletal Fluorosis 64
Soil Erosion 52
Soil Pollution 145
Solar Energy 105
Solar Photovoltaics 108
Solid Phase: Particulate Matter 133
Species 156
Species Composition 13
Species Diversity 77
Step Farming 69
Stockholm Conference 128
Stratification 14
Stratosphere 2
Sulphur Cycle 31
Sustainable Agriculture 49
Sustainable Development 121

T

Terrace Farming 69
Thermal Pollution 142
Thermal pollution 140
Thermosphere 2
Threatened 94
Threatened Species 86
Tidal Energy 110
Toxic Substances 156
Toxicity 156
Trickling Filter 144
Trophic Levels 14
Troposphere 2

U

Urban Disasters 198

V

Value Education 209
Vermicomposting 44, 149

Vermiculture 44

Vulnerable 94

W

Waste Water 143

Wasteland 42

Wasteland Reclamation (Restoration) 70

Water (Prevention and Control of Pollution) Act, 1 169

Water Column 43

Water Cycle 62

Water Pollution 58, 139

Water-borne 63

Watershed Management 197

Wildlife Protection Act, 1972 172

Wind Energy 109

World Summit on Sustainable Development (WSSD) 128

Y

Y-Shaped Energy Flow Model 36

Z

Zooremediation 43

Zoos 93

www.ingramcontent.com/pod-product-compliance
Ingram Content Group UK Ltd.
Pitfield, Milton Keynes, MK11 3LW, UK
UKHW021532300726
14060UKWH00011B/369